AF360745

Advances in In Situ Biological Chemical Groundwater Treatment

Advances in In Situ Biological and Chemical Groundwater Treatment

Editors

Sabrina Saponaro
Snežana Maletić
Elena Sezenna

MDPI • Basel • Beijing • Wuhan • Barcelona • Belgrade • Manchester • Tokyo • Cluj • Tianjin

Editors

Sabrina Saponaro
Politecnico di Milano
Italy

Snežana Maletić
University of Novi Sad
Serbia

Elena Sezenna
Politecnico di Milano
Italy

Editorial Office
MDPI
St. Alban-Anlage 66
4052 Basel, Switzerland

This is a reprint of articles from the Special Issue published online in the open access journal *Water* (ISSN 2073-4441) (available at: https://www.mdpi.com/journal/water/special_issues/groundwater_treatment).

For citation purposes, cite each article independently as indicated on the article page online and as indicated below:

LastName, A.A.; LastName, B.B.; LastName, C.C. Article Title. *Journal Name* **Year**, *Article Number*, Page Range.

ISBN 978-3-03943-432-9 (Hbk)
ISBN 978-3-03943-433-6 (PDF)

Contents

About the Editors

Sabrina Saponaro (Ph.D. in Environmental and Sanitary Engineering) is Associate Professor of Soil Remediation, School of Civil and Environmental Engineering, Department of Civil and Environmental Engineering, Politecnico di Milano (I). She has been working at Politecnico di Milano on sanitary engineering topics since 1996, dealing with fate and transport problems of pollutants in soil and groundwater; health and environmental risk assessment; and traditional and innovative remediation techniques for polluted groundwater, soil, and sediments. A full list of her significant publications can be found at https://orcid.org/0000-0002-9358-6586. Since 2010, she has been scientific coordinator of three projects and local unit head of one project funded on a competitive basis by national and regional funds. She has been a member of the Scientific Committee of the International Symposium on Sediment Management (Ecole des Mines de Douai) since 2010 and the Scientific Committee of RemTech Expo (Ferrara Fiere Congressi) since 2011. She served on the Technical Steering Committee of the 8th International Conference on Remediation and Management of Contaminated Sediments (Battelle) in 2014. Her bibliometric indexes (Scopus) are as follows: 41 publications; 625 citations by 575 papers; h-index 13; 2 patents (as of June 2020).

Snežana Maletić Ph.D. in Chemistry) is Associate Professor of University of Novi Sad, Faculty of Sciences. She has based her career on a multidisciplinary approach, with expertise in environmental science; remediation technology; environmental chemistry; bioavailability and biodegradability of organic contaminants; water, soil, and sediment analysis; and environmental risk assessment. A full list of her significant publications can be found at https:// http://orcid.org/0000-0002-5026-3365. She was coordinator of two projects related to the characterization of biochar and its application as a soil/sediment amendment in order to reduce the bioavailability of toxic organic pollutants. She has participated in 25 national and international projects in the fields of environmental science, environmental risk assessment, and remediation technology. She has been involved in more than 40 studies related to research in environmental protection and technology. Since 2012, she has been Head of the Laboratory for Environmental Chemical Analysis, which is accredited according to ISO 17025 protocols. Her bibliometric indexes (Scopus) are as follows: 46 publications; 450 citations by 412 papers; H-index 13 (as of September 2020).

Elena Sezenna (Ph.D. Environmental and Sanitary Engineering) is Assistant Professor at the Department of Civil and Environmental Engineering at Politecnico di Milano. Her research activity mostly focuses on technologies for remediation of polluted groundwater, soil, and sediments, with special reference to experimental studies of bio-electrochemical processes and electrode-based remediation. The scope of her work extends to fate and transport modeling of contaminants (including emerging pollutants and biodegradable plastics) in soil and groundwater, and she has also been studying human health and environmental risk assessment for the management of contaminated sites, including methods to improve risk estimates and methodologies for the prioritization of remedial efforts. She has been involved as a research participant in various international and national projects since 2003. She is co-author of over 80 publications (21 Scopus-indexed) with 348 citations, and her h-index is 7.

Preface to "Advances in In Situ Biological and Chemical Groundwater Treatment"

Groundwater contamination generically refers to modifications in biological, physical, or chemical characteristics; radioactivity; or the presence of undesirable solutes at significant concentrations. In terms of undesirable solutes, inorganic or organic chemical mixtures frequently occur, including metals and semi-metals, such as chromium and arsenic, and volatile chlorinated hydrocarbons (e.g., tetrachloroethene, trichloroethene).

"Pump and treat" is a common method for cleaning up groundwater contaminated with dissolved chemicals. Groundwater is pumped from wells to an above-ground treatment system that removes the contaminants. Pump and treat may last from a few years to several decades, with the actual cleanup time being long when the concentrations of the contaminants are high, the pollution source has not been completely removed, or the groundwater flow is slow.

The increasing availability of scientific studies has progressively drawn attention to in situ technologies for groundwater remediation. Most of them are innovative compared to the pump-and-treat approach, allowing the remediation time to be reduced and the remediation sustainability to be increased. In situ bioremediation of groundwater involves the encouragement of indigenous bacterial populations to metabolize target contaminants through the addition of various amendments, or the use of selected strains of bacteria in the subsurface to help treatment. Bacteria perform coupled oxidation/reduction reactions to live, and bioremediation exploits all these reactions to remove contaminants from groundwater. Aerobic bioremediation most commonly takes place in the presence of oxygen, and it is most effective in treating non-halogenated organic compounds. Anaerobic reductive bioremediation takes place in the absence of oxygen and promotes the bioreduction of oxidized contaminants such as chlorinated solvents.

Microbes or their enzymes may also effectively remediate toxic heavy metal contamination via their metal-resistance mechanisms, including the transformation of metals into less toxic species, biosorption to the cell wall, entrapment in extracellular capsules, or precipitation.

Nanotechnology is a multidisciplinary field that has gained significant momentum in recent years. The use of nanomaterials, such as zero-valent iron and carbon nanotubes, in the cleanup of groundwater is relatively new and has a great potential for providing efficient, cost-effective, and environmentally acceptable solutions to face the increasing requirements of stringent quality standards. The large surface area of these nanoparticles results in high sorption capacity, along with the ability to be functionalized for the enhancement of their affinity and selectivity. Nano- and microplastics have received widespread attention in recent years as they can sorb various organic contaminants.

Sabrina Saponaro, Snežana Maletić, Elena Sezenna
Editors

Article

Hydrochemical Conditions for Aerobic/Anaerobic Biodegradation of Chlorinated Ethenes—A Multi-Site Assessment

Jan Němeček [1,2,*], Kristýna Marková [2], Roman Špánek [2], Vojtěch Antoš [2], Petr Kozubek [1], Ondřej Lhotský [3,4] and Miroslav Černík [2]

[1] ENACON s.r.o., Krčská 16, CZ-140 00 Prague 4, Czech Republic; kozubek@enacon.cz
[2] Institute for Nanomaterials, Advanced Technologies and Innovation, Technical University of Liberec, Studentská 2, CZ-461 17 Liberec, Czech Republic; kristyna.markova2@tul.cz (K.M.); roman.spanek@tul.cz (R.Š.); vojtech.antos@tul.cz (V.A.); miroslav.cernik@tul.cz (M.Č.)
[3] DEKONTA a.s., Volutová 2523, CZ-158 00 Prague 5, Czech Republic; lhotsky@dekonta.cz
[4] Faculty of Science, Charles University, Benátská 2, CZ-128 01 Prague 2, Czech Republic
* Correspondence: nemecek@enacon.cz or jan.nemecek1@tul.cz

Received: 3 December 2019; Accepted: 19 January 2020; Published: 22 January 2020

Abstract: A stall of *cis*-1,2-DCE and vinyl chloride (VC) is frequently observed during bioremediation of groundwater chloroethenes via reductive dechlorination. These chloroethenes may be oxidised by aerobic methanotrophs or ethenotrophs co-metabolically and/or metabolically. We assessed the potential for such oxidation at 12 sites (49 groundwater samples) using hydrochemical and molecular biological tools. Both ethenotroph (*etnC* and *etnE*) and methanotroph (*mmoX* and *pmoA*) functional genes were identified in 90% of samples, while reductive dehalogenase functional genes (*vcrA* and *bvcA*) were identified in 82%. All functional genes were simultaneously detected in 78% of samples, in actively biostimulated sites in 88% of samples. Correlation analysis revealed that *cis*-1,2-DCE concentration was positively correlated with *vcrA*, *etnC* and *etnE*, while VC concentration was correlated with *etnC*, *etnE*, *vcrA* and *bvcA*. However, feature selection based on random forest classification indicated a significant relationship for the *vcrA* in relation to *cis*-1,2-DCE, and *vcrA*, *bvcA* and *etnE* for VC and no prove of relationship between *cis*-1,2-DCE or VC and the methanotroph functional genes. Analysis of hydrochemical parameters indicated that aerobic oxidation of chloroethenes by ethenotrophs may take place under a range of redox conditions of aquifers and coincide with high ethene and VC concentrations.

Keywords: chlorinated solvents; biological reductive dechlorination; aerobic oxidation; qPCR; ethenotrophs; methanotrophs

---

1. Introduction

Chloroethene tetrachloroethene (PCE) and trichloroethene (TCE) are amongst the most abundant pollutants of groundwater and soil due to their frequent use in industrial applications. These pollutants can be biodegraded through natural or enhanced anaerobic reductive dechlorination, where chloroethenes serve as electron acceptors and molecular hydrogen and acetate, both released as by-products of organic substrate fermentation reactions, are used by the dechlorinating bacteria as electron donors and as carbon sources, respectively [1]. During this process, PCE is converted stepwise to TCE by removing one chlorine atom and replacing it with a hydrogen atom; likewise, trichloroethene (TCE), is primarily converted to *cis*-1,2-DCE, then to vinyl chloride (VC), and finally to ethene [2].

Anaerobic reductive dechlorination of chloroethenes is restricted to just a few bacterial genera (hereafter collectively referred to as anaerobic dechlorinators). Those capable of sequentially dechlorinating PCE or TCE down to *cis*-1,2-DCE include *Dehalobacter* [3,4], *Dehalospirilum* [5], *Desulfuromonas* [6,7], *Geobacter* [8], *Sulfurospirrilium* [9] and *Desulfitobacterium* [10]. *Dehalococcoides mccartyi* [11,12] and *Dehalogenimonas* species [13] are anaerobic dechlorinators known to gain energy through dechlorination of DCE to VC and eventually to ethene using the reductive dehalogenase enzymes BvcA and VcrA [14] or similar ones in the case of *Dehalogenimonas* spp. Despite the presence of anaerobic dechlorinators, *cis*-1,2-DCE and VC often accumulate in groundwater as the sequential steps of the reductive dechlorination process are less and less favourable thermodynamically and kinetically [15], and/or the conditions for complete dechlorination are not always optimal.

Under aerobic conditions, chloroethenes can be oxidised both cometabolically and metabolically. During cometabolic oxidation, chloroethenes are only degraded into non-toxic end-products fortuitously when degrading enzymes are produced for degradation of bacterial growth substrates such as methane, ethene, ammonium or aromatic pollutants. Cometabolic degradation has been shown for all chloroethenes, though only rarely described for PCE [16]. Aerobic cometabolic oxidation is related to certain aerobic bacteria, such as ethene-oxidisers (etheneotrophs) and methane-oxidisers (methanotrophs) [17–19]. Methanotrops employ soluble and particulate methane monooxygenases (sMMO and pMMO, respectively) to oxidise methane as a primary growth substrate. Both sMMO and pMMO are also capable of fortuitous oxidation of chloroethenes. The sMMO have a broader substrate range than pMMO, and are more efficient at degrading chlorinated ethenes [20]. The gene *mmoX*, which encodes the sMMO α subunit, and *pmoA*, which encodes the pMMO α subunit, are used as biomarkers of chloroethene cometabolic potential in groundwater [20–22].

Etheneotrophs can cometabolise VC and DCE when growing on ethene as a primary growth substrate, while several pure etheneotrophic strains, such as *Mycobacterium* and *Nocardiodes*, can also utilise VC as their sole carbon and energy source [18]. Etheneotrophs, when growing on ethene and VC, express a soluble alkene monooxygenase (AkMO), transforming VC to epoxide chlorooxirane, which is further metabolised to 2-chloro-2-hydroxyethyl-CoM by epoxyalkane:coenzyme M transferase (EaCoMT) [19,23]. The genes *etnC* and *etnE* encode the α subunit of AkMO and the EaCoMT, respectively, and serve as emerging biomarkers for ethenotroph-mediated aerobic biodegradation potential, though they do not distinguish between metabolic and cometabolic biodegradation pathways [20,24].

Two degradation pathways have been proposed as regards aerobic metabolic degradation of *cis*-1,2-DCE. Both of them involve degradation through monooxygenase-catalysed epoxidation [25], with the initial step catalysed by cytochrome P450 monooxygenase. Epoxides can be degraded subsequently either by epoxyalkane, coenzyme M transferase or through formation of glutathione conjugates [26,27].

Only a few studies have focused on parallel presence of anaerobic dechlorinators and aerobic methanotrophs or ethenotrophs at contaminated sites. Liang et al. [20] studied the potential for VC degradation at six contaminated sites based on abundance and expression of VC biodegradation genes, and suggested that both ethenotrophs and anaerobic VC dechlorinators simultaneously contributed to VC biodegradation at the sites with high VC attenuation rates. Richards et al. [19] investigated spatial relationships between functional genes of ethenotrophs, anaerobic VC dechlorinators and methanotrophs in aquifer soil samples collected at a contaminated site, and found that functional genes of all the three bacterial guilds coexisted in 48% of the samples that appeared to be anaerobic.

These results attracted our interest to further assess the potential of using the alternate anaerobic/aerobic biodegradation of chloroethenes as a practical remedial tool, which can eliminate frequent accumulation of *cis*-1,2-DCE and VC in groundwater. The goal of this study is to investigate this potential of ongoing activities of both anaerobic and aerobic chloroethene degraders at a large number of remediated sites affected by biostimulation to different levels and to estimate limiting conditions for these microbial degradation processes. For such scanning, the qPCR data and hydrogeochemical parameters were analysed by advanced statistical methods.

2. Materials and Methods

2.1. Test Sites

This study examined 35 sampling wells situated in 16 contaminanted groundwater plumes located at 12 different sites in the Czech Republic (some of the wells were sampled repeatedly, and, thus, 49 groundwater samples were analysed in total). Location and numbering of the sites are depicted in Figure 1. Out of the 16 contaminated groundwater plumes, 12 plumes were remediated using in situ biostimulation of anaerobic biodegradation prior to or over the course of the study. Cheese whey (either in liquid form, as supplied by the Diary Čejetičky spol. s.r.o., Czech Republic, or diluted dry whey supplied by Lacnea agri s.r.o., Czech Republic) was used as an electron donor in all bioremediated plumes. Amounts of whey applied to remedial wells were determined based on the local hydrogeological settings and the target TOC concentration in groundwater >100 mg/L. However, not all the wells in the remediated contaminated groundwater plumes were affected by the applied electron donor. Information on time of whey application to the respective wells (relative to the sampling date) are given in Table 2. Bioaugmentation was not performed at any of the tested sites. On three sites (#2, #3 and #5), zero-valent iron (ZVI) was injected in a pilot scale either alone or together with the whey.

Groundwater samples were collected from shallow aquifers developed mainly in Quaternary fluvial sediments or in a sandy eluvium associated with the crystalline bedrock. Only one contaminant site (#10) was related to a fractured rock aquifer.

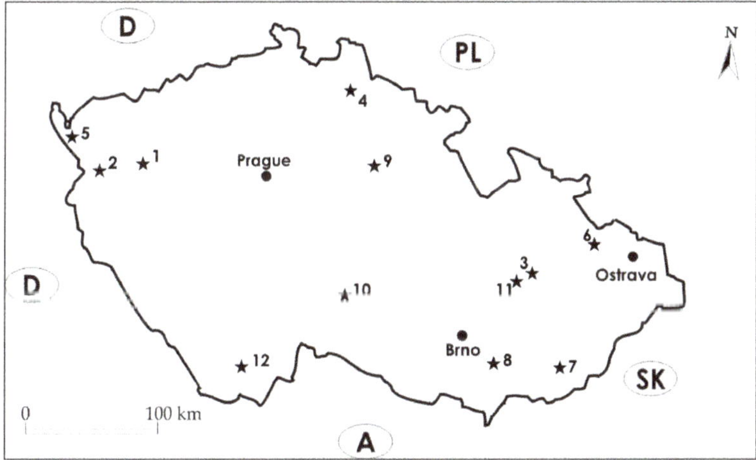

Figure 1. Location and numbering of contaminated sites tested in this study.

2.2. Groundwater Sampling and Scope of Laboratory Analysis

Before sampling, all wells were purged by pumping approximately three borehole casing volumes of groundwater using a Gigant submersible sampling pump (Ekotechnika, Czech Republic), according to standard procedure [28]. A range of field parameters (pH, oxidation-reduction potential (ORP), electrical conductivity and temperature) were recorded using a flow-through cell connected to a Multi 350i Multimeter (WTW, Germany). Samples for real-time PCR analyses were collected into 500 mL single-use, DNA free containers and transported together with the other samples to the laboratory within 24 h.

Groundwater samples from all wells were analysed and monitored for the following parameters: chlorinated ethenes, ethene, ethane, methane, sulfate, hydrogen sulfide, nitrate, dissolved iron and manganese, total organic carbon (TOC), relative abundance of specific bacteria and functional genes using real-time PCR (see Sections 2.3 and 2.4).

2.3. DNA Extraction and Real-Time Quantitative PCR

Groundwater samples (0.2–0.5 L) were filtered through 0.22 µm membrane filters (Merck Millipore, Darmstadt, Germany), after which DNA was extracted from the filters (with microorganisms) using the FastDNA Spin Kit for Soil (MP Biomedicals, Irvine, CA, USA), following the manufacturer's protocol. A Bead Blaster 24 homogenisation unit (Benchmark Scientific, Sayreville, NJ, USA) was employed for cell lysis, while the extracted DNA was quantified using a Qubit 2.0 fluorometer (Life Technologies, Carlsbad, CA, USA).

Real-time quantitative PCR (qPCR) analysis was performed in order to quantify 16S rDNA of total bacteria (16S), the anaerobic dechlorinators *Dehalobacter* spp. (Dre), *D. mccartyi* (Dhc), *Desulfitobacterium* spp. (Dsb) and *Dehalogenimonas* spp. (Dhgm), and genes encoding for reductive dehalogenase (*vcrA* and *bvcA*), functional genes coding enzymes for ethenotroph-mediated aerobic biodegradation (i.e., alkene monooxygenase (*etnC*) and epoxyalkane, coenzyme M transferase (*etnE*)) and functional genes for soluble methane monooxygenase (*mmoX*) and particulate methane monooxygenase (*pmoA*) as biomarkers of methanotroph-mediated aerobic cometabolic biodegradation of chloroethenes. All primers used for qPCR are listed in Supplementary Table S1.

All qPCR assays were performed on a LightCycler® 480 (Roche, Switzerland), using the same reaction conditions described in our previous study [29]. qPCR mixtures with a total volume of 10 µL were prepared using LightCycler® 480 SYBR Green I Master (Roche, Switzerland), 4 pmol of each primer (Generi Biotech, Czech Republic) and 1 µL of template DNA. Each sample was analysed in duplicate in 96-well plates, including no-template controls. The qPCR thermal profile was 5 min at 95 °C followed by 45 cycles at 95 °C for 10 s, 68/60/55 °C for 15 s, and 72 °C for 20 s. Appropriate annealing temperatures are listed in Supplementary Table S1. To control the specificity of the qPCR amplification, a melting curve analysis (72 to 98 °C, ramp rate 0.06 °C/s) was performed at the end of each qPCR. The presence of PCR inhibitors was tested for in each DNA sample by serial dilution of DNA template. Detected Ct values were normalised to the filtration volume and sample dilution to get the final Cq values. qPCR amplification efficiency for each primer set was determined based on the slope of the curves constructed from a serial dilution of template DNA from five different environmental samples. Based on the approach used, the qPCR results were presented as relative abundances of the individual biomarkers.

2.4. Physical and Chemical Parameters of the Groundwater

Concentration of iron and manganese dissolved in groundwater were analysed using an Optima 2100 inductively coupled plasma-optical emission spectrometer (ICP-OES; Perkin Elmer, Waltham, MA, USA) according to ČSN EN ISO 11885 [30]. The groundwater samples were filtered through a 0.45 µm membrane filter prior to analysis. Hydrogen sulfide was determined spectrophotometrically according to ČSN 83 0530-31 [31]. TOC was determined according to ČSN EN 1484 [32] using a MULTI N/C 2100S TOC analyser (Analytik Jena, Jena, Germany). Nitrate and sulfate were assessed using a ICS-90 ion chromatograph (Dionex, Sunnyvale, CA, USA) according to ČSN EN ISO 10304-1 [33]. Volatile organic carbons, including chlorinated ethenes, ethene, ethane and methane, were assessed using a Saturn 2200 CP 3800 gas chromatography–mass spectrometer (GC-MS; Varian, USA) using a VF-624ms column (Varian, Palo Alto, CA, USA), a CTC Combipal injector (CTC Analytics, Morrisville, NC, USA) and a headspace agitator.

2.5. Data Analysis

As an initial step, data below the limit of quantification (LOQ) were replaced with values equal to half of the LOQ of the respective method. Contents of individual chlorinated ethenes, ethene and ethane, in groundwater were converted from mass concentrations to molar concentrations. Dechlorination of the parent chlorinated ethenes PCE and TCE to less chlorinated forms and on to non-chlorinated ethenes through hydrogenolysis was assessed and expressed by the chlorine number (Cl no.), i.e., the weighted average number of chlorine atoms per molecule of ethene [34]. Identification of prevailing redox processes was performed based a range of chemical criteria (Table 1).

Table 1. Water chemistry criteria for identifying redox processes in groundwater (modified from Chapelle et al. [35].

	Predominant Redox Process	NO_3^- (mg/L)	Mn^{2+} (mg/L)	Fe^{2+} (mg/L)	SO_4^{2-} (mg/L)	Fe/H_2S	Methane (mg/L)
Oxic	O_2 reduction	-	<0.05	<0.1	-	-	-
Anoxic	NO_3^- reduction	≥1.0	<0.05	<0.1	-	-	-
	Mn(IV) reduction	<1.0	≥0.05	<0.1	-	-	-
	Fe(III) reduction	<1.0	-	≥0.1	≥0.5	>10	-
	Mix Fe(III)/SO_4^{2-} reduction	<1.0	-	≥0.1	≥0.5	3–10	-
	SO_4^{2-} reduction	<1.0	-	≥0.1	≥0.5	<3	-
	Methanogenesis	<1.0	-	≥0.1	<0.5	-	≥0.5

A final data set, used for further statistical analysis, was created by merging data for hydrochemical parameters and biomarker values.

All statistical analyses were performed in RStudio [36] and R software version 3.6.1 [37]. The relationship between gene abundance and hydrochemical parameters was tested using nonparametric Spearman's correlation [38]. The importance of feature selection attributes was assessed using the Boruta package [39], built on the random forest classification algorithm, which enables a search for significant and non-redundant variables. Biomarker abundance and field parameters were tested for outliers, with positive outliers treated by capping using inter quartile range (IQR = Q3 − Q1, where Q1, Q3 are 1st and 3rd quartile, respectively). Values that lay outside the 1.5 * IQR limits were replaced with 5th percentile. All values were subsequently log-transformed to achieve normality.

The whole dataset was also subjected to a cluster analysis with the ethenotroph functional genes *etnC* and *etnE* set as clustering variables. The dataset included field parameters (pH, ORP), parameters identifying redox processes, i.e., electron acceptors (nitrate, sulfate), and redox reaction products (iron, hydrogen sulfide, and methane), TOC, concentrations of chlorinated ethenes and their non-chlorinated metabolites ethene and ethane, and appropriate biomarkers (*etnC, etnE, mmoX, pmoA, vcrA, bvcA, Dehalobacter* spp., *D. mccartyi, Desulfitobacterium* spp., *Dehalogenimonas* spp.) Six sample clusters were identified representing groundwater samples with the highest and lowest abundance of *etnC* and *etnE*, respectively. For each cluster, basic statistical parameters (maximum, mean, median, and minimum values) were calculated for the hydrochemical parameters. The prevailing redox processes assessed for each groundwater sample (see Section 3.1) were taken into account. Clustering of the final dataset was performed using the hierarchical clustering algorithm implemented by the hclust function in the R software package. The optimal number of clusters and the clustering algorithm were assessed using the clValid package [40].

3. Results and Discussion

3.1. Results of Chemical Analyses

For a full summary of the chemical analyses, see the supplementary material (Supplementary Table S2). As most samples in this study were collected from aquifers affected by historical or on-going remediation using biostimulation via delivery of organic carbon, the laboratory analysis revealed mostly anoxic redox processes. Ongoing methanogenesis was detected based on the applied criteria (≥0.5 mg/L of methane) in 30 of 49 groundwater samples analysed; however, strict methanogenesis was only detected in four samples, the remaining 26 samples displaying criteria for more than one redox process were, mostly including Fe(III) reduction (24 samples). Fe(III) reduction alone, or in combination with Mn(IV) or NO_3^- reduction, was only identified in nine samples.

With regard to concentrations of individual chloroethenes, the prevailing anoxic conditions, favourable for reductive dechlorination, resulted in sequential degradation of the parent contaminants (PCE and/or TCE) down to *cis*-1,2-DCE, VC and ethene. An average Cl no. of 1.5 indicated an advanced state of reductive dechlorination. In 11 of the 49 samples the Cl no. was even below 0.5, showing almost complete dechlorination by biostimulation of anaerobic biodegradation performed at these sites.

In all the collected samples, *cis*-1,2-DCE was the dominant DCE isomer. The ratio of *trans*-1,2-DCE to *cis*-1,2-DCE was below 2.2% in all the samples (mean ratio 0.39%) and concentrations of 1,1-DCE were similar to *trans*-1,2-DCE. Therefore, data for *trans*-1,2-DCE and 1,1-DCE were not included into the final data set for statistical analysis.

Of the 49 samples taken, 43 did not contain any other contaminants (in addition to chloroethenes) in significant concentrations (molar mass of the sum of co-contaminants below 2.5% of the molar mass of the sum of chloroethenes in the respective sample was used as the criterion). Six samples contained significant concentrations of co-occurring contaminants: at contaminated site #6 (two samples), the main contaminants were chloroform and 1,2-dichloroethane, whereas toluene was the dominant contaminant in the samples collected at site #12, although, historically, the site was dominantly contaminated by chloroethenes. Although co-occurring contaminants present in groundwater might affect both anaerobic reductive dechlorination and aerobic biodegradation, they were not included in the final data set for statistical analysis as their occurrence was limited and scattered.

Acetylene as an intermediate of abiotic β-elimination of chloroethenes [41] was detected in seven of the 49 samples taken. It was present at sites #2, #3 and #5 where zerovalent iron materials were injected together with the whey in the past. It can be concluded that abiotic β-elimination has contributed to the degradation of chloroethenes at these sites.

For co-contaminants; trans-1,2-DCE and 1,1-DCE and acetylene concentrations, see the supplementary material (Supplementary Table S3).

3.2. Results of qPCR

qPCR revealed the frequent occurrence of both aerobic and reductive biomarkers. Presence of the ethenotroph functional genes *etnC* and *etnE* was confirmed in 44 of 49 (90%) samples analysed, as were the methanotroph functional genes *mmoX* and *pmoA*, while the reductive dehalogenase genes *vcrA* and *bvcA* were recorded in 40 of 49 (82%) samples analysed (Table 2, Figure 2). All functional genes together (*etnC, etnE, mmoX, pmoA, vcrA* and *bvcA)* were detected in 38 of 49 (78%) samples analysed, indicating that both aerobic oxidation and reductive dechlorination of chloroethenes may take place simultaneously at the same place or in close microenvironments. This finding is consistent with the study of Liang et al. [20], which detected functional genes from all three bacterial guilds (ethenotrophs, methanotrophs and reductive dechlorinators) in 99% of groundwater samples collected at six contaminated sites.

The qPCR results exhibited noticeable differences in individual biomarkers within one site or even one contaminant plume (e.g., aerobic biomarkers in contaminant plume #10_1 or reductive biomarkers in contaminant plume #1_1; see Table 2).

Of the 49 analysed samples, 16 were collected during the on-going remedial biostimulation (samples were collected 1 to 4 months after the last whey application, see column 6 in Table 2). Despite the fact that application of whey stimulates reductive dechlorination, all functional genes (*etnC, etnE, mmoX, pmoA, vcrA* and *bvcA)* coexisted in 14 of the 16 (88%) samples.

Of the anaerobic dechlorinators, *Desulfitobacterium* spp. were most frequent, being present in 47 of 49 samples (96%), while *D. mccartyi* and *Dehalogenimonas* spp. were identified in 46 of 49 analysed samples (94%). The occurrence of *D. mccartyi* and *Dehalogenimonas* spp. corresponded well with *Dehalobacter* spp. (see correlation analysis in Section 3.3), though the latter were only identified in 74% of samples. The lower incidence of *Dehalobacter* spp. in groundwater samples may reflect the fact that

Dehalobacter spp. degrade the parent chlorinated compounds PCE and TCE [3,42], which were mostly degraded to less chlorinated metabolites at the sites tested. Frequent detection of *Dehalogenimonas* spp. in our samples (94%) is consistent with the findings of Yang et al. [13], who detected *Dehalogenimonas* spp. in 81% of 1173 samples collected in the United States and Australia.

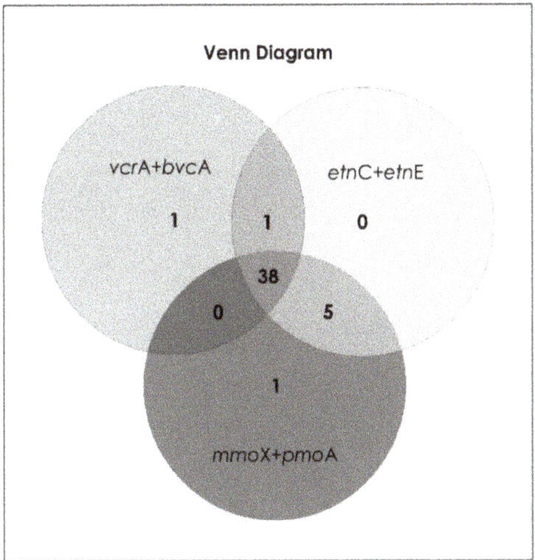

Figure 2. Venn diagram showing the numbers of groundwater samples where functional gene groups were found individually or jointly (value of Cq = 40 was used as the detection limit, see Section 2.3 for more information).

Table 2. Summary of qPCR results (samples are identified based on the well number and time (after underline), site ID refers to Figure 1, plume ID is related to the site ID, and in case that the well was affected by whey application, the time interval is mentioned, for abbreviations see text in Section 2.3).

Sample ID	Site ID	Plume ID	Date of Sampling DD.MM.YY	Affected by Whey Application Y/N	Time Elapsed after the Last Whey Application (Months)	Total Bacteria	Aerobic Biomarkers				VC Reductive Dehalogenase Genes		Reductive Dechlorinators				
						16S	etnC	etnE	mmoX	pmoA	bvcA	vcrA	Dhc	Dhgm	Dsb	Dre
AT-15	1	1_1	24.01.19	Y	1	+++	+	++	+	+	+++	+++	+++	+++	+	+
AT-19	1	1_1	01.05.19	N	NA	+++	++	+++	+	++	++	++	++	++	++	+
AT-20	1	1_1	01.05.19	N	NA	++	++	+++	+	+	ND	+	+	ND	++	ND
AT-21	1	1_1	01.05.19	N	NA	++	++	++	++	++	+-	+	+	++	+	+-
SV-10	2	2_1	24.01.19	Y*	8	+++	++	+++	+	+	+	++	+	+-	+++	ND
VS-5	2	2_1	24.01.19	Y*	8	+++	+++	+++	+	+	+	+	+	+	++	+-
A_I.	3	3_1	04.02.19	N	NA	+	ND	ND	ND	ND	ND	+	+	+	+	ND
B_IV.	3	3_1	04.02.19	N*	NA	+	ND	ND	ND	ND	ND	+	ND	ND	++	ND
SM-7D	3	3_2	04.02.19	Y	16	+++	++	++	+	+	+++	+++	+++	+++	+++	+++
VS-7S_1	3	3_3	18.07.17	N*	NA	+++	+	+	++	++	+	++	+	++	++	+
VS-7S_2	3	3_3	13.10.17	Y	1	+++	+	+	++	+++	+++	+++	+++	+++	++	++
VS-7S_3	3	3_3	12.02.18	Y	4	++	++	++	+	++	+++	+++	+++	+++	++	+
VS-7S_4	3	3_3	26.03.18	Y	5	++	++	++	++	++	+++	+++	+++	+++	++	+
VS-7S_5	3	3_3	04.02.19	Y	16	+++	+	+	+	ND	++	+++	+++	+++	+++	+
Studna_1	3	3_3	18.07.17	N	NA	++	++	++	+	+	++	++	++	++	+	+
Studna_2	3	3_3	13.10.17	Y	1	++	+	+	+	+	+++	+++	+++	+++	++	++
Studna_3	3	3_3	12.02.18	Y	4	+++	+	+	+	+	++	+++	+++	++	+	++
Studna_4	3	3_3	26.03.18	Y	5	+++	+	+	+	+	++	+++	+++	++	+	+
SM-8_1	3	3_3	18.07.17	N	NA	++	++	++	++	++	+++	+++	+++	+++	++	++
SM-8_2	3	3_3	13.10.17	Y	1	+++	++	++	+	+	+++	+++	+++	+++	++	+++
SM-8_3	3	3_3	12.02.18	Y	4	+++	++	++	++	+++	++	+++	+++	+++	++	++
SM-8_4	3	3_3	26.03.18	Y	5	++	++	++	+++	+++	++	+++	+++	++	++	++

Table 2. *Cont.*

Sample ID	Site ID	Plume ID	Date of Sampling DD.MM.YY	Affected by Whey Application Y/N	Time Elapsed after the Last Whey Application (Months)	Total Bacteria	Aerobic Biomarkers				Reductive Biomarkers					
											VC Reductive Dehalogenase Genes		Reductive Dechlorinators			
						16S	*etnC*	*etnE*	*mmoX*	*pmoA*	*bvcA*	*vcrA*	*Dhc*	*Dhgm*	*Dsb*	*Dre*
AP-2	4	4_1	05.02.19	Y	28	++	++	++	++	+	++	++	++	+++	++	ND
HV-16	4	4_1	09.04.19	N	NA	++	+-	+	++	+++	++	++	++	+	+	+-
HV-25	4	4_1	09.04.19	Y	50	++	+	+	++	++	++	++	++	++	+	+
HV-8_1	4	4_1	03.08.16	Y	50	++	+	++	+++	+++	+++	++	++	++	+	+
HV-8_2	4	4_1	10.10.16	Y	1	+++	+	++	+++	+++	+++	+++	+++	++	++	+
HV-8_3	4	4_1	16.01.17	Y	3	++	+	+	+++	+++	++	++	++	+	++	+-
HML-4S_1	4	4_1	03.08.16	Y	50	++	+	+	+	+	++	++	++	+++	++	+
HML-4S_2	4	4_1	10.10.16	Y	1	++	+	+	+	++	++	+++	+++	++	++	+
HML-4S_3	4	4_1	16.01.17	Y	3	++	++	++	++	++	++	++	++	++	++	+
HV-53D	4	4_2	09.04.19	Y	30	++	+	++	+	++	++	++	++	++	++	+
V-5	5	5_1	22.01.19	Y	3	+++	+	++	+	+	++	+++	+++	++	ND	+
V-11	5	5_1	22.01.19	Y	3	+++	+	+	+	+	+	ND	+	++	+++	+
V-13	5	5_2	22.01.19	Y*	3	+++	+	ND	+	+	ND	++	+	+	+++	+
SV-1	5	5_2	22.01.19	N	NA	++	++	+	+	++	++	++	++	++	++	ND
HJ-4	6	6_1	29.01.19	N	NA	++	+	++	+	+	ND	ND	ND	+-	+	ND
V-32	6	6_1	29.01.19	N	NA	+++	++	++	+	+	ND	+	+-	+	++	ND
MV - 6A	7	7_1	20.02.19	Y	34	+	+-	+-	+	+	+	+	+	++	+	ND
Z - 4	7	7_1	20.02.19	Y	34	++	++	++	++	++	++	++	++	++	+	ND
Žd - 2	8	8_1	20.02.19	Y	26	+++	+++	+++	+++	+++	++	+++	+++	+++	++	++
Žd - 4	8	8_1	20.02.19	Y	26	+++	+++	+++	++	+++	+++	+++	+++	++	+++	+++
MR 4	9	9_1	21.02.19	Y	50	++	+	+	++	+	++	++	++	++	++	+

Table 2. *Cont.*

Sample ID	Site ID	Plume ID	Date of Sampling DD.MM.YY	Affected by Whey Application Y/N	Time Elapsed after the Last Whey Application (Months)	Total Bacteria	Aerobic Biomarkers				Reductive Biomarkers					
											VC Reductive Dehalogenase Genes		Reductive Dechlorinators			
						16S	etnC	etnE	mmoX	pmoA	bvcA	vcrA	Dhc	Dhgm	Dsb	Dre
ZMS 4	9	9_1	21.02.19	Y	9	++	+	++	+++	+++	++	++	++	++	++	ND
HLV-5	10	10_1	04.03.19	N	NA	+++	+	++	+++	++	ND	++	++	++	++	+
HV-26	10	10_1	04.03.19	N	NA	+	ND	ND	ND	ND	ND	ND	ND	ND	+	ND
ID-2	11	11_1	05.03.19	N	NA	+	ND	ND	ND	ND	+	+	+	+	ND	ND
HV-223	12	12_1	09.05.19	Y	2	+++	+	++	+++	+++	+++	+++	+++	++	++	++
HV-112	12	12_1	09.05.19	Y	2	+++	+	+-	+++	+++	+++	++	++	++	++	+

Legend: * affected by application of ZVI

+++	high quantity
++	medium quantity
+	low quantity
+-	close to the detection limit
ND	not detected

Categories of "high", "medium", and "low" quantitites represent the respective thirds of the determined range of Cq values for each biomarker

3.3. Correlation and Feature Selection Analysis

Correlation analysis (Figure 3) revealed that TCE was only correlated with *Desulfitobacterium* spp., probably as none of the enzymes tested related to the reductive or aerobic functional genes participating in biodegradation of TCE. However, *cis*-1,2-DCE was positively correlated with both the reductive functional gene *vcrA* and the ethenotroph functional genes *etnC* and *etnE*, as well as the reductive dechlorinators *Dehalogenimonas* spp. and *Desulfitobacterium* spp. Similarly, VC concentration was positively correlated to the ethenotroph functional genes *etnC* and *etnE* and the reductive dehalogenase genes *vcrA* and *bvcA*, as well as the reductive dechlorinators *D. mccartyi* and *Dehalogenimonas* spp. The positive correlation of *cis*-1,2-DCE and VC to abundance of *Dehalogenimonas* spp., whose ability to degrade *cis*-1,2-DCE and VC was only recently confirmed by Yang et al. [13], indicates a potentially notable contribution of *Dehalogenimonas* spp. in reductive dechlorination at the sites tested.

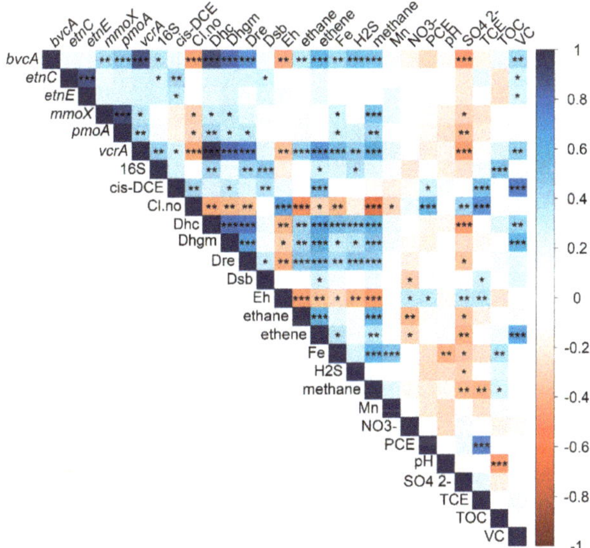

Figure 3. Results of Spearman correlation analysis (intensities of blue and red colours indicate values of positive and negative correlation coefficients, respectively; *p*-values: *** = 0.001, ** = 0.01, * = 0.05).

None of the chlorinated ethenes were correlated with the methanotroph functional genes *mmoX* and *pmoA*.

In sum, it implies that both anaerobic reductive dechlorination and ethenotroph-mediated aerobic biodegradation could participate in biodegradation of VC and *cis*-1,2-DCE.

With regard to the correlation of individual functional genes with hydrochemical and field parameters, the reductive dehalogenase genes *vcrA* and *bvcA* were negatively correlated with ORP and sulfate and positively correlated with dissolved iron, hydrogen sulfide and methane. As has been previously noted, anaerobic reductive dechlorination of chloroethenes can occur under both nitrate-reducing and iron-reducing conditions, though the most favourable conditions for anaerobic dechlorinators are sulfate reducing and methanogenic [43]. This is also supported by a strong negative correlation of Cl no. with methane and positive correlation of Cl no. with sulfate. The methanotroph functional genes *mmoX* and *pmoA* were positively correlated with dissolved iron and methane, while *pmoA* was negatively correlated with sulpfate. There was no correlation observed between the ethenotroph functional genes *etnC* and *etnE* and any hydrochemical and/or field parameters, suggesting that hydrochemical conditions of aquifers are not limiting factors for proliferation of ethenotrophs.

The significance of individual *cis*-1,2-DCE and VC biodegradation processes was assessed using feature selection utilising the random forest classification algorithm. Only the reductive dehalogenase gene *vcrA* exhibited significant relevance for *cis*-1,2-DCE (Figure 4), whereas both reductive dehalogenase genes (*vcrA* and *bvcA*) and the ethenotroph functional gene *etnE* were significantly relevant for VC (Figure 5). In sum, feature selection provided further support for the hypothesis that both reductive dechlorinators and ethenotrophs participate in biodegradation of VC at the sites tested; however, it indicated that there was only intermediate significant involvement of ethenotrophs in biodegradation of *cis*-1,2-DCE.

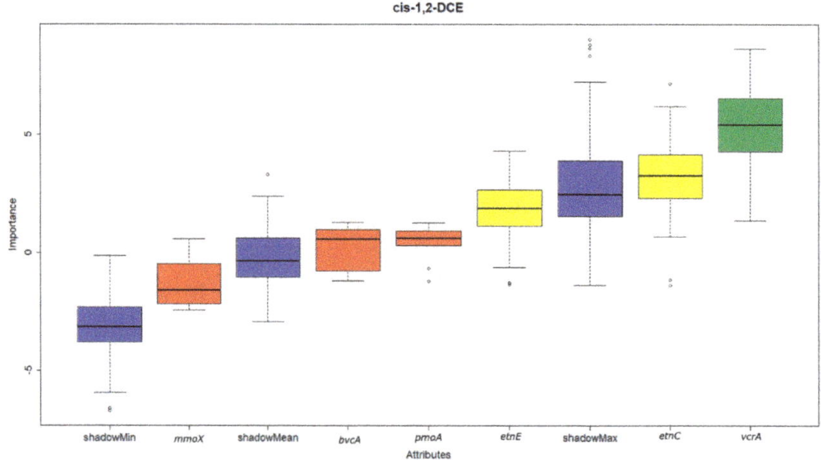

Figure 4. Importance of individual biomarkers on concentration of *cis*-1,2-DCE, based on the Boruta algorithm utilising the random forest classification. Green indicates significant attributes, yellow indicates attributes of intermediate significance while red indicates attributes that are not significant (blue colour indicates shadows—Boruta auxiliary features used for selection of significant attributes).

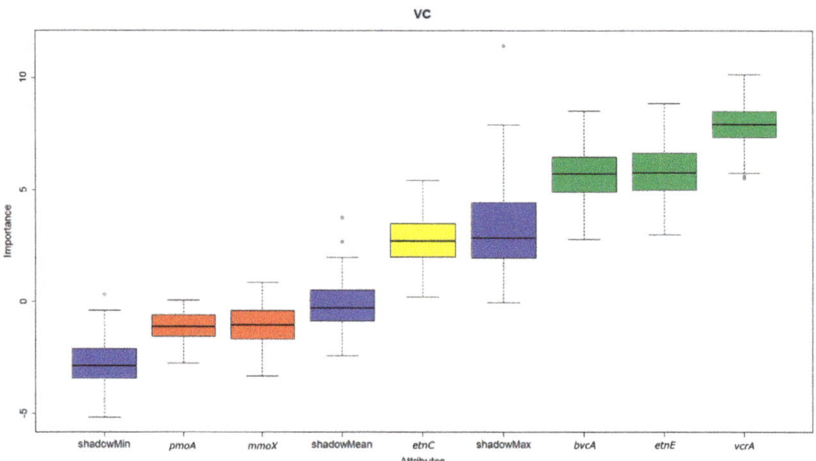

Figure 5. Importance of individual biomarkers on concentration of VC, based on the Boruta algorithm utilising the random forest classification. Green indicates significant attributes, yellow indicates attributes of intermediate significance while red indicates attributes that are not significant (blue colour indicates shadows—Boruta auxiliary features used for selection of significant attributes).

3.4. Hydrochemical Conditions for Aerobic Oxidation

As aerobic (both metabolic and co-metabolic) oxidation processes in remedial practice can overcome a frequent accumulation of metabolites *cis*-1,2-DCE and VC generated during in-situ anaerobic reductive bioremediation, the hydrochemical conditions for these processes were assessed. First, the whole dataset was subjected to cluster analysis with the ethenotroph functional genes *etnC* and *etnE* set as clustering variables. Six sample clusters were identified, with clusters #1 and #6 representing groundwater samples with the highest and lowest abundance of *etnC* and *etnE*, respectively. The basic statistical parameters (maximum, mean, median, and minimum values) calculated for the hydrochemical parameters for each cluster are given in graphs in the supplementary material (Supplementary Figure S1). The prevailing redox processes assessed for each groundwater sample (see Section 3.1) are given in Supplementary Table S2.

Cluster #1, which represented the highest potential for aerobic oxidation, was comprised of three groundwater samples collected from two sites. Redox conditions were mixed, covering a wide spectrum of processes from Fe(III) reduction (iron mean concentration 33.38 mg/L) to methanogenic (methane mean concentration 7.8 mg/L). The groundwater samples in cluster #1 differed from those in the other clusters, mainly in higher concentrations of VC (mean 10,314 µg/L) and of ethene (mean 4534 µg/L).

In comparison, cluster #6 comprised groundwater samples with lowest potential for aerobic oxidation and included 28 samples collected at eight sites. With regards to redox conditions, cluster #6 contained samples with Mn(VI) reduction and Fe(III) reduction predominating, along with samples of mixed redox categories, covering Mn(VI) reduction down to methanogenic conditions. Mean iron and methane concentrations for cluster #6 were 15.8 and 3.5 mg/L, respectively. Groundwater samples in this cluster contained significantly lower concentrations of VC (mean 276 µg/L) and ethene (mean 344 µg/L); however, the degree of chloroethene dechlorination (Cl no.) was similar to that of cluster #1 (mean Cl no. cluster #1 = 1.5, cluster #6 = 1.4). This indicates that aerobic oxidation of chloroethenes can take place in aquifers under a range of redox conditions, including apparently reducing environments. The most likely explanation is that the studied aquifers are not homogenous environments with one redox state only, but highly spatially and temporally heterogenous macro- and microenvironments with different redox conditions. Spatial heterogeneities may be a result of different lithology and permeability that influence migration patterns of organic substrates (electron donors), dissolved oxygen, chloroethenes and their co-occurring contaminants. This explanation is supported by the results of hotspot high-resolution characterisation performed on sites #2, #3, #5, #6, #7, #8 and 11 using a membrane interface probe (MIP) (Geoprobe Systems®, Salina, KS, USA). MIP was used for collection of semi-quantitative data on the presence of volatile organic compounds (VOCs) and soil electric conductivity measurements indicating aquifer lithology in the vertical soil profile. MIP profiles (data not shown) from majority of the sites show presence of soil layers with different electric conductivities and levels of contamination (i.e., distinct lithological and contamination heterogeneity).

Temporal changes may result from irregular seepage of oxic rainwater, groundwater level fluctuations, and/or from discontinuous application of organic substrates at the bioremediated sites. Methanotrophs are found mainly at aerobic/anaerobic interfaces in soil and aquatic environments that are crossed by methane [44], thus inhabiting environments with low oxygen level. Similarly, ethenotrophs can survive in environments with very limited oxygen contents [17,45,46]. These macro- and microheterogeneities and abilities of aerobic ethenotrophs to sustain low levels of oxygen are reasons for their occurrence in anaerobic aquifers containing anaerobic dechlorinators. Co-occurrence of anaerobic dechlorinators and VC assimilating bacteria was revealed also in groundwater samples from other sites [20,29], in discrete aquifer soil samples [19] as well as in surface riverbed sediment samples [47] where the TOC content in soil was found to be the critical parameter determining the dominant degradation pathway [48].

The results also suggest that high concentrations of ethene and VC (as electron donors in aerobic oxidation of chloroethenes) are correlated with the abundance of ethenotrophs. This is also supported by the results of correlation analysis, which indicated a positive correlation between VC and the ethenotroph functional genes *etnC* and *etnE*.

4. Conclusions

The aim of this study is to assess the potential for aerobic oxidation of chlorinated ethenes at contaminated sites and to characterise those conditions favourable for aerobic biodegradation of VC and *cis*-1,2-DCE.

The main findings of this study are as follows:

1. Both the ethenotroph functional genes *etnC* and *etnE* and methanotroph functional genes *mmoX* and *pmoA* were identified in 90% of groundwater samples, while all functional genes (*etnC, etnE, mmoX, pmoA, vcrA* and *bvcA)* coexisted in 78% of samples, in actively biostimulated sites in 88% of samples.
2. The reductive dechlorinator *Dehalogenimonas* spp., only recently identified as capable of *cis*-1,2-DCE and VC degradation, was detected by qPCR in 94% of samples. A positive correlation between *Dehalogenimonas* spp. abundance and *cis*-1,2-DCE and VC concentration indicates a potential contribution to reductive dechlorination at the sites tested.
3. Presence of both *cis*-1,2-DCE and VC was positively correlated with the reductive functional gene *vcrA*, and the ethenotroph functional genes *etnC* and *etnE*, and VC additionally with the reductive functional gene *bvcA*.
4. Of all the functional genes tested, only the reductive dehalogenase functional gene *vcrA* was found to be significant for *cis*-1,2-DCE degradation by feature selection using the random forest algorithm. On the hand, both the dehalogenase functional genes *vcrA* and *bvcA* and the ethenotroph functional gene *etnE* were indicated as significant for VC.
5. No significant relationship was observed between *cis*-1,2-DCE and VC concentration and abundance of the methanotroph functional genes *mmoX* and *pmoA*, because methanotrophs oxidise these contaminants cometabolically only, without any energetic benefit.
6. Cluster analysis revealed that aerobic oxidation of chloroethenes can take place under a broad range of apparent redox conditions, even under apparently methanogenic conditions, probably due to a high redox microheterogeneity of the aquifer environment, ability of ethenotrophs to survive in environments with very limited oxygen contents and high concentrations of ethene and VC (as electron donors in aerobic oxidation of chloroethenes), these being the most important hydrochemical parameters affecting abundance of ethenotrophs.

The results of this study demonstrate the potential for incorporating aerobic steps in in-situ bioremediation schemes for treatment of chloroethenes, especially when temporal stalls in carcinogenic VC may cause an on-site human health risk.

Supplementary Materials: The following are available online at http://www.mdpi.com/2073-4441/12/2/322/s1, Table S1: Primers used in qPCR, Table S2: Hydrochemical data and assessment of predominant redox processes, Table S3: Concentrations of volatile organic compounds and acetylene in groundwater, Figure S1: Graphs of hydrochemical and molecular biological data for each cluster of groundwater samples.

Author Contributions: Conceptualization, J.N.; methodology, J.N. and K.M.; Investigation, P.K. and O.L.; Laboratory analyses, V.A. and K.M.; data curation, J.N., R.Š. and K.M.; Writing—original draft preparation, J.N.; Writing—review and editing, M.Č., K.M. and O.L. All authors have read and agreed to the published version of the manuscript.

Funding: This research was funded by the Technology Agency of the Czech Republic, grant number TH04030225.

Acknowledgments: The authors acknowledge the assistance provided by Research Infrastructure NanoEnviCz (Project no. LM2015073), supported by the Ministry of Education, Youth and Sports of the Czech Republic.

Conflicts of Interest: The authors declare no conflict of interest.

References

1. Bradley, P.M.; Chapelle, F.H. *In Situ Remediation of Chlorinated Solvent Plumes*; Stroo, H., Ward, C., Eds.; Springer: New York, NY, USA, 2010; Chapter 3; pp. 39–67. ISBN 978-1-4419-1400-2.

2. Stroo, H.F.; West, M.R.; Kueper, B.H.; Borden, R.C.; Major, D.W.; Ward, C.H. *Chlorinated Solvent Source Zone Remediation*; Kueper, B.H., Stroo, H.F., Vogel, C.M., Ward, C.H., Eds.; Springer: New York, NY, USA, 2014; Chapter 12; pp. 395–457. ISBN 978-1-4614-6921-6.

3. Holliger, C.; Hahn, D.; Harmsen, H.; Ludwig, W.; Schumacher, W.; Tindall, B.; Vazquez, F.; Weiss, N.; Zehnder, A.J. *Dehalobacter restrictus* gen. nov. and sp. nov., a strictly anaerobic bacterium that reductively dechlorinates tetra-and trichloroethene in an anaerobic respiration. *Arch. Microbiol.* **1998**, *169*, 313–321. [CrossRef] [PubMed]

4. Holliger, C.; Schraa, G.; Stams, A.; Zehnder, A. A highly purified enrichment culture couples the reductive dechlorination of tetrachloroethene to growth. *Appl. Environ. Microbiol.* **1993**, *59*, 2991–2997. [CrossRef]

5. Neumann, A.; Scholz-Muramatsu, H.; Diekert, G. Tetrachloroethene metabolism of *Dehalospirillum multivorans*. *Arch. Microbiol.* **1994**, *162*, 295–301. [CrossRef] [PubMed]

6. Krumholz, L.R.; Sharp, R.; Fishbain, S.S. A freshwater anaerobe coupling acetate oxidation to tetrachloroethylene dehalogenation. *Appl. Environ. Microbiol.* **1996**, *62*, 4108–4113. [CrossRef] [PubMed]

7. Sung, Y.; Ritalahti, K.M.; Sanford, R.A.; Urbance, J.W.; Flynn, S.J.; Tiedje, J.M.; Löffler, F.E. Characterization of two tetrachloroethene-reducing, acetate-oxidizing anaerobic bacteria and their description as *Desulfuromonas michiganensis* sp. nov. *Appl. Environ. Microbiol.* **2003**, *69*, 2964–2974. [CrossRef] [PubMed]

8. Sung, Y.; Fletcher, K.E.; Ritalahti, K.M.; Apkarian, R.P.; Ramos-Hernández, N.; Sanford, R.A.; Mesbah, N.M.; Löffler, F.E. *Geobacter lovleyi* sp. nov. strain SZ, a novel metal-reducing and tetrachloroethene-dechlorinating bacterium. *Appl. Environ. Microbiol.* **2006**, *72*, 2775–2782. [CrossRef] [PubMed]

9. Luijten, M.L.; de Weert, J.; Smidt, H.; Boschker, H.T.; de Vos, W.M.; Schraa, G.; Stams, A.J. Description of *Sulfurospirillum halorespirans* sp. nov., an anaerobic, tetrachloroethene-respiring bacterium, and transfer of *Dehalospirillum multivorans* to the genus *Sulfurospirillum* as *Sulfurospirillum multivorans* comb. nov. *Int. J. Syst. Evol. Microbiol.* **2003**, *53*, 787–793. [CrossRef]

10. Maillard, J.; Regeard, C.; Holliger, C. Isolation and characterization of Tn-Dha1, a transposon containing the tetrachloroethene reductive dehalogenase of *Desulfitobacterium hafniense* strain TCE1. *Environ. Microbiol.* **2005**, *7*, 107–117. [CrossRef]

11. Löffler, F.E.; Ritalahti, K.M.; Zinder, S.H. *Bioaugmentation for Groundwater Remediation*; Stroo, H., Leeson, A., Ward, C., Eds.; Springer: New York, NY, USA, 2013; Chapter 2; pp. 39–88. ISBN 978-1-4614-4114-4.

12. Maymó-Gatell, X.; Nijenhuis, I.; Zinder, S.H. Reductive dechlorination of cis-1, 2-dichloroethene and vinyl chloride by "*Dehalococcoides ethenogenes*". *Environ. Sci. Technol.* **2001**, *35*, 516–521. [CrossRef]

13. Yang, Y.; Higgins, S.A.; Yan, J.; Şimşir, B.; Chourney, K.; Iyer, R.; Hettlich, R.L.; Baldwin, B.; Ogles, D.M.; Löffler, F.E. Grape pomace compost harbors organohalide-respiring *Dehalogenimonas* species with novel reductive dehalogenase genes. *ISME J.* **2017**, *11*, 2767–2780. [CrossRef]

14. Judger, B.-E.; Ertan, H.; Lee, M.; Manefield, M.; Marquis, C.P. Reductive dehalogenases come of age in biological destruction of organohalides. *Trends Biotechnol.* **2015**, *33*, 596–608. [CrossRef]

15. Tratnyek, P.G.; Johnson, R.L.; Lowry, G.V.; Brown, R.A. *Chlorinated Solvent Source Zone Remediation*; Kueper, B.H., Stroo, H.F., Vogel, C.M., Ward, C.H., Eds.; Springer: New York, NY, USA, 2014; Chapter 10; pp. 307–351. ISBN 978-1-4614-6921-6.

16. Thiem, A.; Schmidt, K.R. Sequential anaerobic biodegradation of chloroethenes—Aspects of field application. *Curr. Opin. Biotechnol.* **2011**, *22*, 415–421. [CrossRef] [PubMed]

17. Coleman, N.V.; Mattes, T.E.; Gossett, J.M.; Spain, J.C. Phylogenetic and kinetic diversity of aerobic vinyl chloride-assimilating bacteria from contaminated sites. *Appl. Environ. Microbiol.* **2002**, *68*, 6162–6171. [CrossRef] [PubMed]

18. Mattes, T.E.; Alexander, A.K.; Coleman, N.V. Aerobic biodegradation of the chloroethenes: Pathways, enzymes, ecology, and evolution. *FEMS Microbiol. Rev.* **2010**, *34*, 445–475. [CrossRef]

19. Richards, P.M.; Liang, Y.; Johnson, R.; Mattes, T.E. Cryogenic soil coring reveals coexistence of aerobic and anaerobic vinyl chloride degrading bacteria in a chlorinated ethene contaminated aquifer. *Water Res.* **2019**, *157*, 281–291. [CrossRef]

20. Liang, Y.; Liu, X.; Singletary, M.A.; Wang, K.; Mattes, T.E. Relationships between the abundance and expression of functional genes from vinyl chloride (VC)-degrading bacteria and geochemical parameters at VC-contaminated sites. *Environ. Sci. Technol.* **2017**, *51*, 12164–12174. [CrossRef]

21. Wymore, R.A.; Lee, M.H.; Keener, W.K.; Miller, A.R.; Colwell, F.S.; Watwood, M.E.; Sorenson, K.S. Field evidence for intrinsic aerobic chlorinated cometabolism by methanotrophs expressing soluble methane monooxygenase. *Biorem. J.* **2007**, *11*, 125–139. [CrossRef]

22. Paszynski, A.J.; Paidisetti, R.; Johnson, A.K.; Crawford, R.L.; Colwell, F.S.; Green, T.; Delwiche, M.; Lee, H.; Newby, D.; Brodie, E.L.; et al. Proteomic and targeted qPCR analyses of ubsurface microbial communities for presence of methane monooxygenase. *Biodegradation* **2011**, *22*, 1045–1059. [CrossRef]

23. Coleman, N.V.; Spain, J.C. Epoxyalkane: Coenzyme M transferase in the ethene and vinyl chloride biodegradation pathways of mycobacterium strain JS60. *J. Bacteriol.* **2003**, *185*, 5536–5545. [CrossRef]

24. Jin, Y.O.; Mattes, T.E. A quantitative PCR assay for aerobic, vinyl chloride- and ethene-assimilating microorganisms in groundwater. *Environ. Sci. Technol.* **2010**, *44*, 9036–9041. [CrossRef]

25. Abe, Y.; Aravena, R.; Zopfi, J.; Shouakar-Stash, O.; Cox, E.; Roberts, J.D.; Hunkeler, D. Carbon and chlorine isotope fractionation during aerobic oxidation and reductive dechlorination of vinyl chloride and cis-1, 2-dichloroethene. *Environ. Sci. Technol.* **2009**, *43*, 101–107. [CrossRef] [PubMed]

26. Jennings, L.K.; Chartrand, M.M.; Lacrampe-Couloume, G.; Lollar, B.S.; Spain, J.C.; Gossett, J.M. Proteomic and transcriptomic analyses reveal genes upregulated by cis-dichloroethene in *Polaromonas* sp. strain JS666. *Appl. Environ. Microbiol.* **2009**, *75*, 3733–3744. [CrossRef] [PubMed]

27. Dolinová, I.; Štrojsová, M.; Černík, M.; Němeček, J.; Macháčková, M.; Ševcǔ, A. Microbial degradation of chloroethenes: A review. *Environ. Sci. Pollut. Res.* **2017**, *24*, 13262–13283. [CrossRef] [PubMed]

28. ČSN ISO 5667-11. *Water Quality—Sampling—Part 11: Guidance on Sampling of Groundwaters*; International Organization for Standardization: Geneva, Switzerland, 1993.

29. Němeček, J.; Dolinová, I.; Macháčková, J.; Špánek, R.; Ševcǔ, A.; Lederer, T.; Černík, M. Stratification of Chlorinated Ethenes Natural Attenuation in an Alluvial Aquifer Assessed by Hydrochemical and Biomolecular Tools. *Chemosphere* **2017**, *184*, 1157–1167. [CrossRef]

30. ČSN EN ISO 11885. *Water Quality—Determination of Selected Elements by Inductively Coupled Plasma Optical Emission Spectrometry (ICP-OES)*; International Organization for Standardization: Geneva, Switzerland, 2007.

31. ČSN 83 0530-31. *Chemical and Physical Analysis of Surface Water. Determination of Sulfide and Hydrogen Sulfide*; International Organization for Standardization: Geneva, Switzerland, 1980.

32. ČSN EN 1484. *Water AnalysisGuidelines for the Determination of Total Organic Carbon (TOC) and Dissolved Organic Carbon (DOC)*; International Organization for Standardization: Geneva, Switzerland, 1998.

33. ČSN EN ISO 10304-1. *Water Quality—Determination of Dissolved Anions by Liquid Chromatography of Ions—Part 1: Determination of Bromide, Chloride, Fluoride, Nitrate, Nitrite, Phosphate and Sulfate*; International Organization for Standardization: Geneva, Switzerland, 2007.

34. Bewley, R.; Hick, P.; Rawcliffe, A. Meeting the challenges for bioremediation of chlorinated solvents at operational sites: A comparison of case studies. In Proceeding of the 13th International UFZ-Deltarez Conference on Sustainable Use and Management of Soil, Sediment and Water Resources, Copenhagen, Denmark, 9–12 June 2015.

35. Chapelle, F.H.; Bradley, P.M.; Thomas, M.A.; McMahon, P.B. Distinguishing iron-reducing from sulfate-reducing conditions. *Ground Water* **2009**, *47*, 300–3058. [CrossRef]

36. RStudio Team. *RStudio: Integrated Development for R*; RStudio, Inc.: Boston, MA, USA, 2015; Available online: http://www.rstudio.com/ (accessed on 28 November 2019).

37. R Core Team. *R: A Language and Environment for Statistical Computing*; R Foundation for Statistical Computing: Vienna, Austria, 2018; Available online: http://www.R-project.org/ (accessed on 28 November 2019).

38. Myers, J.L.; Well, A.D. *Research Design and Statistical Analysis*, 2nd ed.; Lawrence Erlbaum Associates: Mahwah, NJ, USA, 2003; p. 760. ISBN 0805840370.

39. Kursa, M.B.; Rudnicki, W.R. Feature Selection with the Boruta Package. *J. Stat. Soft.* **2010**, *36*, 1–13. [CrossRef]

40. Brock, E.; Pihur, V.; Datta, S.; Datta, S. clValid: An R Package for Cluster Validation. *J. Stat. Soft.* **2008**, *25*, 1–22. [CrossRef]

41. Roberts, A.L.; Totten, L.A.; Arnold, W.A.; Burris, D.R.; Campbell, T.J. Reductive elimination of chlorinated ethylenes by zero valent metals. *Environ. Sci. Technol.* **1996**, *30*, 2654–2659. [CrossRef]

42. Wild, A.; Hermann, R.; Leisinger, T. Isolation of an anaerobic bacterium which reductively dechlorinates tetrachloroethene and trichloroethene. *Biodegradation* **1996**, *7*, 507–511. [CrossRef]

43. Bouwer, E.J. Bioremediation of chlorinated solvents using alternate electron acceptors. In *Handbook of Bioremediation*; Norris, R., Hinchee, R., Brown, R., McCarty, P., Semprini, L., Wilson, J., Kampbell, D., Reinhard, M., Borden, R., Eds.; Lewis Publishers: Boca Raton, FL, USA, 1994; pp. 149–175.

44. Ehrlich, H.L.; Newman, D.K. *Geomicrobiology*, 5th ed.; CRC Press, Taylor & Francis Group: Boca Raton, FL, USA, 2009; p. 628. ISBN 978-0-8493-7906-2.

45. Schmidt, K.R.; Tiehm, A. Natural attenuation of chloroethenes: Identification of sequential reductive/oxidative biodegradation by microcosm studies. *Water Sci. Technol.* **2008**, *58*, 1137–1145. [CrossRef]

46. Gossett, J.M. Sustained aerobic oxidation of vinyl chloride at low oxygen concentrations. *Environ. Sci. Technol.* **2010**, *44*, 1405–1411. [CrossRef] [PubMed]

47. Atashgahi, S.; Maphosa, F.; Doğan, E.; Smidt, H.; Springael, D.; Dejonghe, W. Small-scale oxygen distribution determines the vinyl chloride biodegradation pathway in surficial sediments of riverbed hyporheic zones. *FEMS Microbiol. Ecol.* **2013**, *84*, 133–142. [CrossRef] [PubMed]

48. Atashgahi, S.; Lu, Y.; Ramiro-Garcia, J.; Peng, P.; Maphosa, F.; Sipkema, D.; Dejonghe, W.; Smidt, H.; Springael, D. Geochemical parameters and reductive dechlorination determine aerobic cometabolic vs. aerobic metabolic vinyl chloride biodegradation at oxic/anoxic interface of hyporheic zones. *Environ. Sci. Technol.* **2017**, *51*, 1626–1634. [CrossRef] [PubMed]

Article

Microbial Assisted Hexavalent Chromium Removal in Bioelectrochemical Systems

Gabriele Beretta [1,*], **Matteo Daghio** [2,3], **Anna Espinoza Tofalos** [2], **Andrea Franzetti** [2], **Andrea Filippo Mastorgio** [1], **Sabrina Saponaro** [1] and **Elena Sezenna** [1]

[1] Department of Civil and Environmental Engineering, Politecnico di Milano, Piazza Leonardo da Vinci 32, 20133 Milano, Italy; andreafilippo.mastorgio@polimi.it (A.F.M.); sabrina.saponaro@polimi.it (S.S.); elena.sezenna@polimi.it (E.S.)

[2] Department of Earth and Environmental Sciences, University of Milano-Bicocca, Piazza della Scienza 1, 20126 Milano, Italy; matteo.daghio@unimib.it (M.D.); anna.espinoza@unimib.it (A.E.T.); andrea.franzetti@unimib.it (A.F.)

[3] Department of Agriculture, Food, Environment and Forestry, University of Florence, Piazzale delle Cascine 18, 50144 Firenze, Italy

* Correspondence: gabriele.beretta@polimi.it; Tel.: +39-02-23996435

Received: 28 November 2019; Accepted: 5 February 2020; Published: 10 February 2020

Abstract: Groundwater is the environmental matrix that is most frequently affected by anthropogenic hexavalent chromium contamination. Due to its carcinogenicity, Cr(VI) has to be removed, using environmental-friendly and economically sustainable remediation technologies. BioElectrochemical Systems (BESs), applied to bioremediation, thereby offering a promising alternative to traditional bioremediation techniques, without affecting the natural groundwater conditions. Some bacterial families are capable of oxidizing and/or reducing a solid electrode obtaining an energetic advantage for their own growth. In the present study, we assessed the possibility of stimulating bioelectrochemical reduction of Cr(VI) in a dual-chamber polarized system using an electrode as the sole energy source. To develop an electroactive microbial community three electrodes were, at first, inserted into the anodic compartment of a dual-chamber microbial fuel cell, and inoculated with sludge from an anaerobic digester. After a period of acclimation, one electrode was transferred into a polarized system and it was fixed at −0.3 V (versus standard hydrogen electrode, SHE), to promote the reduction of 1000 µg Cr(VI) L^{-1}. A second electrode, served for the set-up of an open circuit control, operated in parallel. Cr(VI) dissolved concentration was analysed at the initial, during the experiment and final time by spectrophotometric method. Initial and final microbial characterization of the communities enriched in polarized system and open circuit control was performed by 16S rRNA gene sequencing. The bioelectrode set at −0.3 V showed high Cr(VI) removal efficiency (up to 93%) and about 150 µg L^{-1} day^{-1} removal rate. Similar efficiency was observed in the open circuit (OC) even at about half rate. Whereas, purely electrochemical reduction, limited to 35%, due to neutral operating conditions. These results suggest that bioelectrochemical Cr(VI) removal by polarized electrode offers a promising new and sustainable approach to the treatment of groundwater Cr(VI) plumes, deserving further research.

Keywords: bioelectrochemical systems (BESs); hexavalent chromium; electrobioremediation; groundwater treatment

---

1. Introduction

Extensive use of chromium (Cr) and its compounds in many industrial process [1] and refractory production has made it a major pollutant [2]. The mobility, bioavailability and toxicity of chromium depend on its oxidation states. In the natural environment, Cr(III) is most immobile, less soluble and stable. Whereas, Cr(VI) is highly mobile, soluble and bioavailable. Compared with Cr(III), Cr(VI) is also extremely toxic (by 100 times) to living organisms and it is internationally recognized as a human carcinogen, mutagen and teratogen.

To reduce its impact on human health and the environment, appropriate remedial measures and remediation interventions are needed. The recovery and removal of chromium from wastewater or groundwater is traditionally carried out by adsorption, [3,4], chemical or electrochemical reduction to Cr(III), and subsequent precipitation [5,6] or electrokinetics [7]. These methods bring with them some disadvantages that concern high-energy requirements, excessive chemical consumption, production of unwanted secondary products and residual highly concentrated toxic sludge [1,5]. Furthermore, some of these approaches are quite effective at the high chromium concentrations [3,7] of industrial effluents. Whereas, the removal efficiency greatly decreases at low concentrations, as typically observed in natural surface and groundwater.

Bioremediation, exploiting the huge microbial metabolic capacity to transform contaminants into harmless substances, may overcome some of the drawbacks of physical-chemical technologies, and is a low energy and cost-effective process. Microbes naturally adopt different strategies to survive in chromium polluted environment, such as biosorption, bioaccumulation and biotransformation for detoxification of Cr(VI) into the relatively safe Cr(III) form [8]. Bioremediation of Cr(VI) includes all these mechanisms and may be promoted through the addition of nutrients and/or electron donors to sustain microbial growth (biostimulation), or even by the injection of selected bacterial strains that are able to reduce Cr(VI) and enhance the removal process (bioaugmentation).

Recently, great attention has been also paid to Microbial Electrochemical Technology (MET) as an innovative and sustainable approach for promoting the bioremediation of contaminated sites [9]. In METs, microorganisms catalyze the oxidation or reduction of pollutants using solid-state electrodes as virtually inexhaustible electron acceptors or donors [10–12].

MET has achieved promising results in the treatment of wastewater contaminated with Cr(VI). In microbial fuel cells (MFCs), a representative of MET, Cr(VI) in industrial effluents could be effectively reduced in either, abiotic cathodic chambers, under acidic conditions [13–16] or biocathodes in the neutral pH range [17,18], while generating electricity from low-grade anodic organic substrates. Cr(VI) can also be removed in systems relying on an external power supply for creating a bias between two electrodes or a potentiostat to supplement electrons to the cathode [19,20].

Under anaerobic conditions, Cr(VI) can serve as the final electron acceptor in a process that usually involves membrane-bound reductases [21], but also soluble enzymes (e.g., c3 cytochrome in Desulfovibrio v.) [22]. Previous work has confirmed Cr(VI) autotrophic reduction in soil-aquifer that has involved different species of microorganisms (e.g., *Clostridium chromiireducens* sp., *Pseudomonas Synxantha*) that are able to use hydrogen and CO_2/$NaHCO_3$ as an electron donor, and carbon source, respectively [23–26]. In the wide variety of bacteria capable of efficient Cr(VI) reduction [27] some, belonging to the species *Shewanella* sp. [20], *Desulfovibrio* sp. [22], *Pseudomonas* sp. [28], *Trichococcus* sp. [18], *Stenotrophomonas* sp., *Serratia* sp., and *Achromobacter* sp. [29] have been in fact reported to be able to use an electrode as electron donor and Cr(VI) as electron acceptor.

The advantages to exploiting the ability of these microorganisms to bioremediate ensures issues in electron donors delivery and availability to be avoided, and improved control of parasitic reactions to avoid the formation of unwanted daughter products [9]. Moreover, the production of an electrical signal can act as a biosensor for real-time monitoring of the microbial activity [30,31].

This work aimed to determine the effectiveness of a microbial biocathode for CrVI reduction in comparison to a pure microbial and pure electrochemical control, to start exploring the possibility of effective Cr(VI) reduction in natural surface water or groundwater at lower Cr(VI) concentrations than wastewater so far investigated. Therefore, an initial concentration of Cr(VI) of 1 mg L^{-1} was applied and no organic substance was added during the chromium removal phase. To minimize the total energy required and to facilitate the rapid development of an electroactive microbial biofilm, we divided our work into two phases. In the first phase, electrodes were acclimated in the anodic chamber of MFC, inoculated with anaerobic digester sludge. Once the electroactive biofilm has developed, one of the electrodes was transferred in a dual-chamber potentiostatically-controlled system (POL −0.3 V) and −0.3 V versus Standard Hydrogen Electrode (SHE) potential was imposed (unless otherwise stated, all potentials throughout the paper are relative to SHE). To approach real groundwater conditions, differently from previous studies [32–34], no organic substance was added in the working chamber, dosing carbonates as the sole carbon source.

2. Materials and Methods

2.1. Reactors Set-Up and Operation

For this study, five identical double-chamber ("H-shaped") reactors have been set up (Figure 1). Each reactor consisted of a pair of borosilicate-glass bottles, 1.2 L useful volume each one. The connection between anolyte and catholyte was achieved by a proton exchange membrane (PEM, 4.52 cm^2, Nafion117, FuelCellsEtc, College Station, TX, USA) placed between bottles. Graphite cylinders (ATAL Grafiti, Trezzo sull'Adda, Italy, length 6 cm, diameter 1 cm, geometric area 18.85 cm^2) served as electrodes, both anodes and cathodes. Stainless steel cables (Ø 0.1 cm) fixed in the centre of the graphite cylinders were used as current collectors. All cables were covered with heat-shrinkable polytetrafluoroethylene (PTFE) tubes (Sigma-Aldrich, Milan, Italy) to limit corrosion phenomena. The distance between anode and cathode within the system was about 10 cm. In the experiments, a mineral medium (7 g L^{-1} NaH$_2$PO$_4$ · 12H$_2$O, 3 g L^{-1} KH$_2$PO$_4$, 1 g L^{-1} NH$_4$Cl and 0.5 g L^{-1} NaCl), pre-autoclaved at 120 °C for 30 min twice, was used as electrolyte and to support microbial growth. For the entire duration of the test, all the systems were kept at a constant temperature and pH, respectively, 18 ± 1 °C and 7.4 ± 0.1.

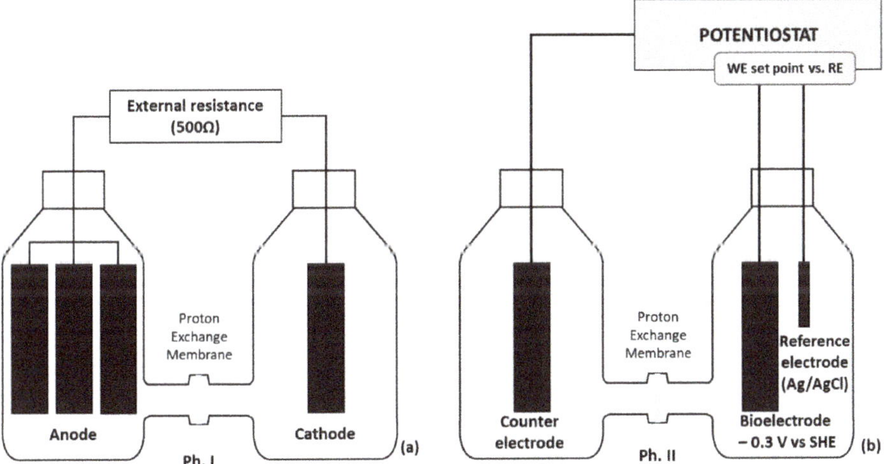

Figure 1. Schematic diagram of the reactors used in the present study. (**a**) In the acclimatization phase (Ph. I) of the electroactive biofilm (MFC), anode and cathode were connected by a 500 Ω external resistance. (**b**) In the second phase (Ph. II), in the polarized reactor (POL −0.3 V) and in the abiotic polarized control (ABI −0.3 V), a potential of −0.3 V versus SHE was imposed.

2.2. Electroactive Biofilm Development in MFC

In the first phase of the work, a MFC was set up in one of the previously described reactors, in order to develop anodic biofilm able to use the anode as the main electrons acceptor (Figure 1a). Three electrodes were placed in the anode chamber in parallel, to simultaneously develop an electroactive biofilm on all three electrodes, and one in the cathodic chamber. The anodic chamber was inoculated with anaerobic digester sludge (0.24 L, corresponding to 20% of the volume) and then sodium acetate was added, up to a concentration of 0.1 g L^{-1}, as the carbon source for microbial growth. To establish anaerobic conditions, during the set-up, the anodic solution was flushed with nitrogen and the chamber immediately sealed with screw caps fitted with PTFE/silicone gaskets. The cathode chamber, filled with mineral medium, was kept open to air to allow oxygen diffuse into the solution.

For the entire duration of the test, in the external circuit connecting anode and cathode, a constant resistance (500 Ω) was maintained and voltage drop continuously recorded (PicoLog 1012, Pico Technology Ltd., Eaton Socon, UK). This allowed the produced current density to be monitored over time, and to spike further sodium acetate into the anodic compartment, as soon as current density below 0.5 mA m^{-2} was recorded.

In this first phase, for the characterization of the microbial community, two samples were taken: A sample from anaerobic sludge, used as an inoculum in MFC (t0 Ph. I), and a sample of the anodic solution, used to inoculate the potentiostatically-controlled tests in the following experimental phase (t0 Ph. II).

2.3. Polarized Bio-Electrode for Cr(VI) Removal

Two out three graphite cylinders, acclimated in the anodic chamber of the MFC, were used to set up (i) the polarized system at −0.3 V versus SHE for the removal of Cr(VI) (POL −0.3 V), and (ii) an open circuit control (OC). The configuration of the systems is the same as described above (Figure 1b). The acclimatized bio-electrode was immersed in the polarized system working chamber filled with mineral medium, KHCO$_3$ (2g L^{-1}) as the sole source of carbon, and K$_2$Cr$_2$O$_7$ (Cr(VI) 1 mg L^{-1}). The working chamber housed an Ag/AgCl reference electrode (0.2 V versus SHE). In addition to the biofilm on the electrode, the same chamber was inoculated with 0.24 L of solution from the anodic chamber of the MFC. To minimize the amount of dissolved organic carbon, the inoculum was first subjected to a washing procedure, by repeated centrifugation (10 min at 4000 rpm, Thermo Scientific) and resuspension of the pellet in fresh mineral medium. The counter electrode chamber was filled up with the same volume of mineral medium and KHCO$_3$ as used for the working chamber. POL −0.3 V has been connected to a dual-channel potentiostat controlled by an Arduino board (Politecnico di Milano I3N-DICA, 2016 Cariplo-BEvERAGE) and a potential of −0.3 V was imposed to the bioelectrode. As described in [35,36], the potentiostat-board allows setting potential to the bioelectrode, that act as working electrode (WE), at set point (i.e., −0.3 V versus SHE) against an Ag/AgCl reference electrode (+0.2 V versus SHE) placed in the same chamber of the WE. The third electrode acts as counter electrode (CE) and potentiostat adjusts the current flow in CE to maintain the fixed potential to the WE.

The OC system, to assess the biological Cr(VI) reduction, was set up exactly like POL −0.3 V without connecting the external circuit. Two abiotic control reactors were also set up (i) ABI −0.3 V in which a new electrode was polarized at −0.3 V, to evaluate electrochemical Cr(VI) removal and (ii) ABI-OC, in which no potential was imposed to assess any absorption phenomena. The configuration of the abiotic controls was the same as the POL −0.3 V. The solutions were sterilized in an autoclave twice (120 °C for 30 min) before filling up abiotic controls.

Cr(VI) removal test POL −0.3 V and OC lasted respectively 6 and 12 days, until more than 90% chromium removal was reached. Whereas abiotic tests (ABI −0.3 V and ABI-OC) lasted 12–14 days until no further changes in chromium concentrations were observed. Through these different systems, it was possible to compare the reduction of Cr(VI) exclusively by the electrochemical, biological and bio-electrochemical way.

The different phases of the experimental work, the acronyms used and their descriptions are summarized in Table 1 and Figure 2.

Table 1. Summary table of the experimental work.

Work Phase	Description	Acronym	Carbon Source	Hexavalent Chromium	Microbial Characterization
Electroactive bacteria enrichment	Inoculum in MFC	t0 Ph. I	Organic matter (anaerobic sludge)	-	Planktonic community
	MFC until 15th day—inoculum of POL −0.3 V and OC	t0 Ph. II	Acetate	-	
Cr(VI) removal tests	Potentiostatically controlled system	POL −0.3 V	KHCO$_3$	K$_2$Cr$_2$O$_7$ (1 mg L^{-1} Cr(VI))	Planktonic community and biofilm developed on polarized electrode (POL −0.3 V electrode)
	Open circuit control	OC			Planktonic community and biofilm developed on graphite (OC graphite)
	Potentiostatically controlled abiotic system	ABI −0.3 V			-
	Open circuit abiotic control	ABI-OC			-

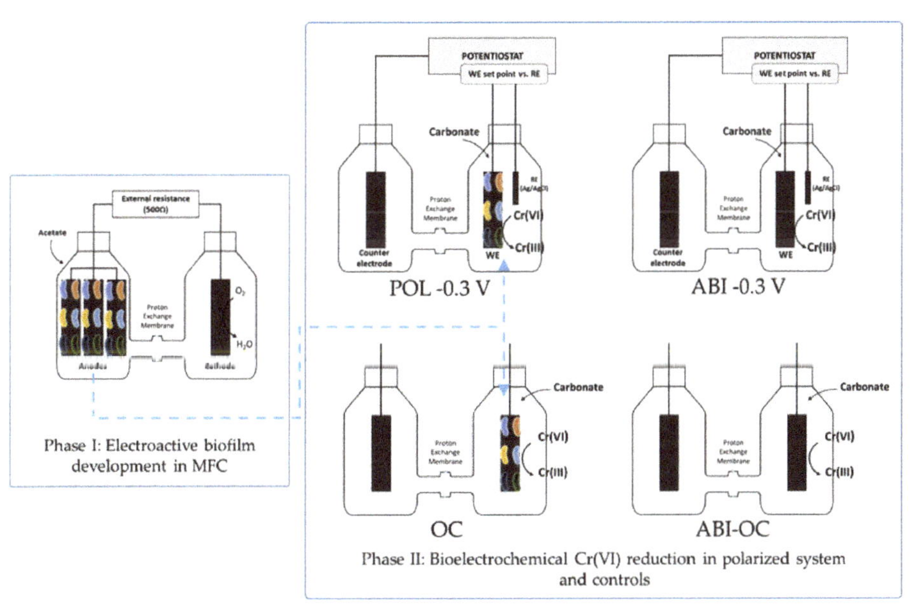

Figure 2. Conceptual model of the experimental work.

2.4. Analyses and Data Processing

The samples were periodically collected from all the systems for the analysis of Cr(VI) by means of a spectrophotometer (DR 6000, Hach Company, Loveland, CO, USA), according to the APHA Standard Methods 3500—B method [37]. Detection limit of spectrophotometric method was 18 µg Cr(VI) L^{-1}. Only Cr(VI) was monitored because of its soluble form and because the risks for human health associated with the trivalent forms of dissolved chromium are much lower, compared to Cr(VI).

To characterize the microbial communities, (i) samples of the anaerobic sludge, the anodic solution of the MFC and the solutions of OC and POL −0.3 V were filtered on 0.45 µm sterile paper filters and

(ii) the biofilms attached to the POL −0.3 V electrode and OC graphite were scraped with a sterile scalpel obtaining graphite powder and biofilm. The genomic DNA was extracted using the FastDNA Spin Kit for Soil (MP Biomedicals, Solon, OH, USA) according to the manufacturer's instructions. The V5-V6 hypervariable regions of the 16S rRNA gene were PCR-amplified using the 783F and 1046R primers [38,39]. The bacterial PCR was performed in 20 µL volume reactions with GoTaq® Green Master Mix (Promega Corporation, Madison, WI, USA) and 1 µM of each primer. After the amplification, DNA quality was evaluated spectrophotometrically and DNA was quantified using Qubit® (Life Technologies, Carlsbad, CA, USA). The sequencing was carried out at Consorzio per il Centro di Biomedicina Molecolare (CBM). Bioinformatics elaborations have been performed as previously reported [40]. Classification of the representative sequences of each Operational Taxonomic Unit (OTU) was done using the RDP classifier (≥80% confidence) [41]. Rarefaction curves, OTU richness, Shannon and Chao1 alpha-diversity indices were generated for each sample using R (vegan).

2.5. Calculations

The MFC current was calculated as $I = V/R$ (A), where V (V) is the voltage drop across the external resistance, R (in Ω). The current density (mA m^{-2}) was calculated as $J = I/AE$ where AE is the geometric area of the electrode (m^2). Coulombic efficiency (EC) is the percentage of electrons circulating in the system, compared to the total electrons theoretically present in the oxidizable organic substrate (Equation (1)). The EC of the MFC, therefore, indicates the microbial conversion efficiency of the substrate into electric current. EC was calculated by assuming the dosed acetate as the only oxidizable substrate and sole electron source in the reactors,

$$EC = \frac{(MI\Delta t)}{(FbV_{an}\Delta S)} \tag{1}$$

where M (59 g mol^{-1}) is the molecular weight of CH_3COO^-, I (A) the recorded current in time Δt (s). F is the Faraday constant (96,485.3 Coulomb mol^{-1} of electrons), b is the stoichiometric factor = 8 moles of electrons per mole of acetate, V_{an} (L) is the volume of the anodic compartment in the MFC and ΔS the consumed substrate over time (g L^{-1}). The consumed substrate was calculated by assuming the complete oxidation of acetate recorded current was below 0.5 mA m^{-2}.

Chromium removal efficiency was calculated by considering the percentage residual dissolved Cr(VI) concentration with time, or,

$$Residual\ Cr(VI) = \frac{C(t)}{C_0} \times 100 \tag{2}$$

where C_0 (µg L^{-1}) and $C(t)$ (µg L^{-1}) are the initial concentration and concentration at time t (days), respectively. Cr(VI) removal efficiency with time is therefore:

$$Cr(VI)\ removal\ efficiency = \left(1 - \frac{C(t)}{C_0}\right) \times 100 \tag{3}$$

3. Results

3.1. Current Density in the MFC

The current density in the MFC, throughout the experiment, varied from a minimum of 0.5 mA m^{-2} to a maximum of 237 mA m^{-2} (Figure 3). At the beginning of the test, from the set-up (day 0) to the first spike of acetate (day 3), the current density and the EC reached the peak values of, respectively, 160 mA m^{-2}, and 2.5%. Following the third addition of acetate (day 7), both an increase in current density (max ~ 230 mA m^{-2}) and in EC (18%) was observed. Even after the fourth addition of acetate, the response of the system in terms of current density was the same for previous spikes (max ~ 230 mA m^{-2}), while the EC value (16.5%) was slightly reduced. The decrease in EC was probably due to the cathodic reaction [42] limiting the overall process, and thus, favouring acetate consumption by different metabolic pathways (acetoclastic methanogenesis) instead of the anode-using ones [43].

After 15 days from the MFC set-up, two out three electrodes and 0.24 L of solution from the anodic chamber were removed for the set-up of POL −0.3 V and open-circuit systems. The removed anolyte was replaced by an equal volume of mineral medium and acetate (0.1 g L^{-1}). The current density in the cycles, following renewal, immediately reached the same peak values recorded in the previous cycles, confirming the cathodic reaction as the limiting step in the whole process. The EC, vice versa, gradually increased up to about 14%. The trend in both the currents and the ECs, before, and after, electrode transfer and the renewal of the medium, shows that the microbial community from the initial inoculum, following an adaptation phase, was able to develop a biofilm able to oxidize the acetate and to transfer electrons to the anode with a discrete efficiency.

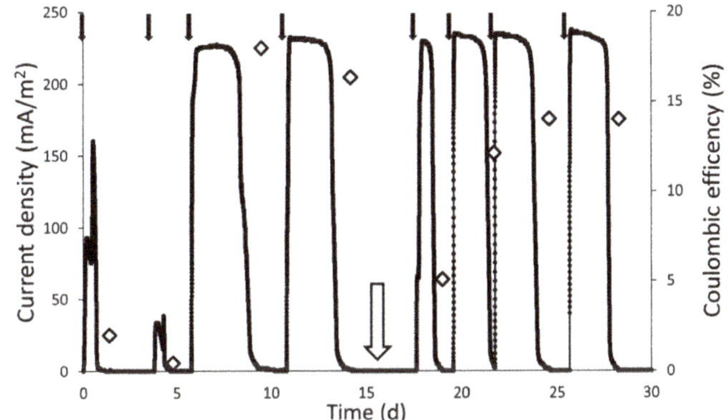

Figure 3. Current density and Coulombic Efficiency trend. Current density (black line) produced by electroactive biofilm after periodic acetate addition (black arrows) until the solution was renewed (white arrow) and two of the three electrodes used to set up POL −0.3 V and OC. The EC was calculated (empty diamond) at the end of each current peak.

3.2. Cr(VI) trends

Biological Cr(VI) reduction, either in the POL −0.3 V and OC, outstanded the abiotic reduction. In the POL −0.3 V system, a rapid decrease in the concentration of dissolved hexavalent chromium was observed (Figure 4). After 6 days from test set-up, the residual concentration in POL −0.3 V was equal to 92.1 ± 13.2 µg L^{-1}, corresponding to a removal rate of about 151 µg L^{-1} day^{-1}. A slower removal rate, about 75 µg L^{-1} day^{-1}, approximately half the POL −0.3 V, was registered in the the OC test. However, the overall removal was relevant, with low residual Cr(VI) concentration at the end of the test of 24.6 ± 7.7 µg L^{-1}.

In both the bioreduction tests, POL −0.3 V and OC, Cr(VI) reduction rate was not influenced by a declining in the Cr(VI) concentration. We hypothesise that, in our biological tests, Cr(VI) could be used as a terminal electron acceptor [44,45]. The higher removal rate in the polarized system might be ascribed to a larger array of electron donors, with respect to the open circuit control. In addition to endogenous electron reserves, the electrode posed at −0.3 V can act as a further electron source [46], that could have promoted chromium reduction. Additional biological mechanisms of chromium reduction, such as enzyme reduction (soluble or membrane-associated reductase) [21], intracellular detoxification mechanisms [45,47] and adsorption on the cell surface [48,49] may have occurred equally in both biological systems.

ABI-OC showed adsorption onto materials did not provide any significant contribution to the removal of Cr(VI), with an overall efficiency not exceeding 3% in the operational conditions of the test.

In ABI −0.3 V about 35% removal was recorded over 12 days. Following an initial rapid decline, in the first 3 days, about 68.5 µg L^{-1} day^{-1}, Cr(VI) removal rate slowed down, leading in about a week to an asymptotic residual concentration of 728 ± 35 µg L^{-1}, corresponding to 31 µg L^{-1} day^{-1} average removal rate. This purely electrochemical Cr(VI) reduction appears to be somehow limited by the operating conditions, neutral pH (7.4 ± 0.1) and poised working electrode potential, as already pointed out, even at higher initial Cr(VI) concentrations, in previous experiences [50,51]. Moreover, as already reported in other studies, Cr(III) precipitation could passivate the electrode further limiting Cr(VI) reduction [14].

By comparing Cr(VI) removal rates in the different tests, it is possible to observe how, in our experimental conditions, biolelectrochemical reduction in POL −0.3 V, about 150 µg L^{-1} Cr(VI), appears to be about 20% faster than the simple superimposition of purely electrochemical and bioreduction mechanisms (31.4 µg L^{-1} day^{-1} in ABI −0.3 V and 75 µg L^{-1} in OC).

Due to the presence of an electroactive biofilm in this study, it was possible to observe more than 90% hexavalent chromium reduction at neutral pH. Conversely, in previous studies, the electrochemical reduction of Cr(VI) was found to be strongly dependent on the pH of the cathodic solution [13,15]. Singhvi and Chhabra demonstrated low pH increases the reduction rate of Cr(VI) to Cr(III) and the concomitant production of electrical energy in abiotic cathodes, while Gangadharan and Nambi did not observe any reduction in a cathodic solution at pH 7 [50,52].

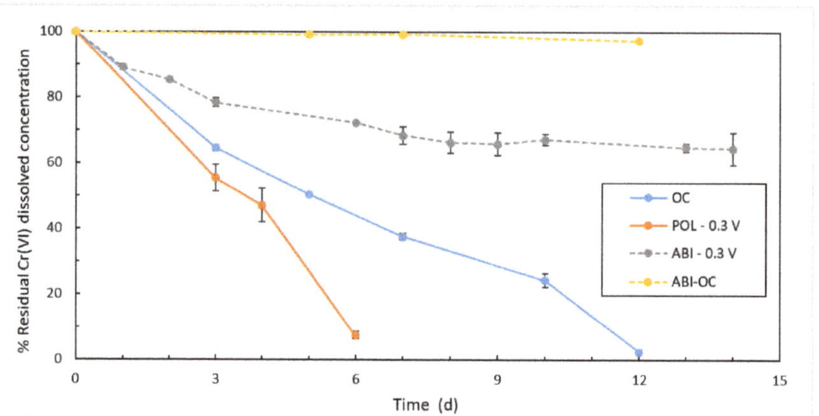

Figure 4. Reduction of Cr(VI) concentrations over time in ABI-OC (yellow dashed line), OC (blue solid line), POL −0.3 V (orange solid line), ABI −0.3 V (pink dashed line).

3.3. Microbial Communities

From the four samples, a total of 98,719 amplicon reads were obtained. Rarefaction curves showed an appreciable representation of the real OTU richness for four samples (t0 Ph. II, OC, POL −0.3 V and POL −0.3 V electrode) (Figure S1). Whereas, the observed OTU richness is probably an underestimation of the real richness of samples t0 Ph. I and OC graphite. To overcome this underestimation, the Chao1 index was calculated, in order to estimate the exact richness [53]. Data analysis shows that the diversity of the microbial communities decreased on the electrodes compared with the bulks. The decrease in diversity was shown by both a decrease in the Shannon and Chao1 indices from the bulks (Shannon: ranging from 4.33 for POL −0.3 V to 4.43 for OC; Chao1: ranging from 333.23 for t0 Ph. I to 635.19 for POL −0.3 V), compared to the electrodes (Shannon: 4.14 for OC graphite and 3.68 for POL −0.3 V electrode; Chao1: 343.77 for OC graphite and 437.00 for POL −0.3 V electrode). These decreases in diversity likely resulted from the selection of electroactive microorganisms on the electrodes (Table S1).

The results of the microbiological analyses showed an evolution bacterial community to begin from the inoculum (t0 Ph. I) to the inoculum of POL −0.3 V and OC (t0 Ph. II) (Figure 5). The microbial communities show an increase in the relative abundance of bacteria belonging to the *Burkholderiales* (from <1.5 to 25.5%), *Bacteroidales* (from 3.1 to 11.7%), *Pseudomonadales* (from <1.5 to 9.4%) and *Clostridiales* (from 10.4 to 12.6%) orders. The representatives of the *Burkholderiales* (*Alcaligenaceae, Comamonadaceae*), *Bacteroidales* (*Porphyromonadaceae*), *Pseudomonadales* (*Pseudomonadaceae*) orders have been previously described as bacteria able to perform electrochemical interactions with the anode [54–58]. Likewise, the presence of the families *Porphyromonadaceae, Comamonadaceae* and *Pseudomonadaceae* were described on the cathode [59] or in cathodic chambers [60].

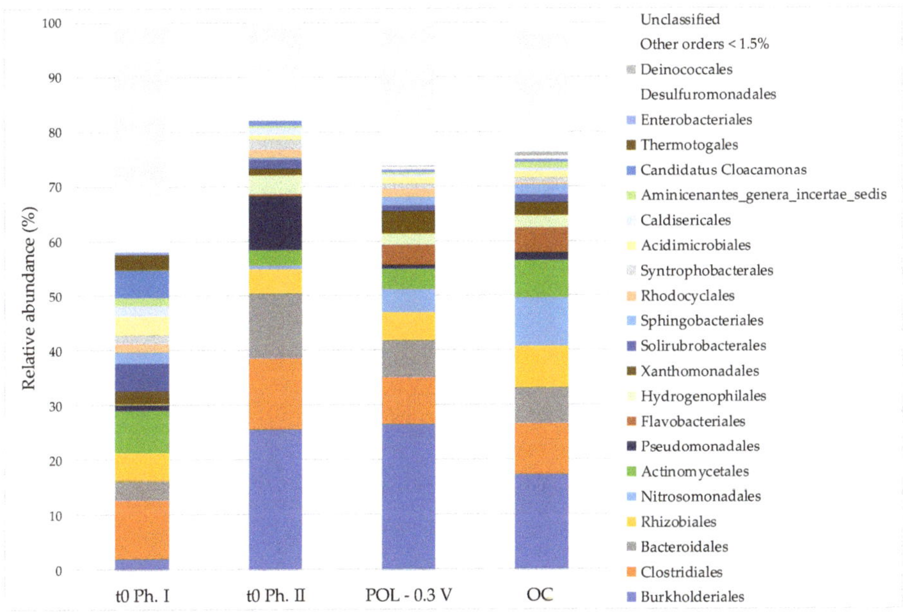

Figure 5. Planktonic bacterial communities structure at order level in the inoculum of MFC (t0 Ph. I), beginning of the second phase (t0 Ph. II), polarized system (POL −0.3 V) and open circuit (OC). "Other orders <1.5%" indicated strains with abundance less than 1.5%.

The addition of Cr(VI) and carbonates as the sole carbon source have influenced the structure of the bacterial community from the beginning of the second phase (t0 Ph. II). Compared to the inoculum, both in POL −0.3 V and in OC systems, *Pseudomonadales, Bacteroidales* and *Clostridiales* reduced their relative abundances by about 80%, 50% and 30%, respectively. Viceversa, in OC, an increase in the relative abundance of the *Nitrosomonadales,* from <1.5% to 8.8%, was observed. On the other hand, in POL −0.3 V there was a slight increase in the orders *Burkholderiales* due to a net increase in the population of *Comamonadaceae,* from 2.8% in t0 Ph. II to 10.1%. In the same system also relative abundance of *Xanthomonadales* increased respect to the inoculum. Both the POL −0.3 V and the OC showed relative abundance increase, compared to the inoculum of the bacteria belonging to the *Flavobacteriales* order, from 0.3% to 3.6%, and 4.4%, respectively.

The lack of organic electron donors, such as acetate, has influenced the structure of the suspended bacterial community, suggesting a possible advantage for autotrophic microorganisms (Nitrosomonadales and Flavobacteriales) able to use carbonates and ammonium in the mineral medium as carbon source and inorganic electron donor [61]. In the POL −0.3 V solution, the presence of the polarized electrode may have favored bacteria belonging to *Burkholderiales* order (*Comamonadaceae* and *Alcaligenaceae* families), as already reported in other studies as components of the electroactive communities, enriched from anaerobic digester sludge and chromium tolerant—resistant/autotrophic microrganisms [62–64].

Furthermore, bacteria belonging to these two families have been observed in bioelectrochemical systems, both for the removal of inorganic compounds, such as H_2S and NO_3^- under autotrophic or mixotrophic conditions [63,65], and for Cr(VI) bioreduction [21,66].

The results of 16S rRNA gene sequencing of the biofilms developed on the polarized electrode (POL −0.3 V electrode) and on the graphite in the open circuit control (OC graphite) were compared, starting from the order level (Figure 6) and it was deepened on a genus level (Table 2). The *Burkholderiales* Order, observed in both the solutions of the inoculated systems (17.4% in OC and 26.8% in POL −0.3 V) and in OC graphite (22.1%), is present instead with a relative minor abundance on POL −0.3 V electrode (16.5%). On the polarized bioelectrode bacteria belonging to the orders *Flavobacteriales* (16.9% compared to 4.1% in OC graphite), *Rhizobiales* (13.7% compared to 5% in OC graphite), and *Deinococcales* (4.9% compared to 1.7% in OC graphite) were more abundant; instead the same were almost absent in the planktonic communities (*Flavobacteriales* 3.6–4.4%, *Rhizobiales* <1.5%, *Deinococcales* <1.5%).

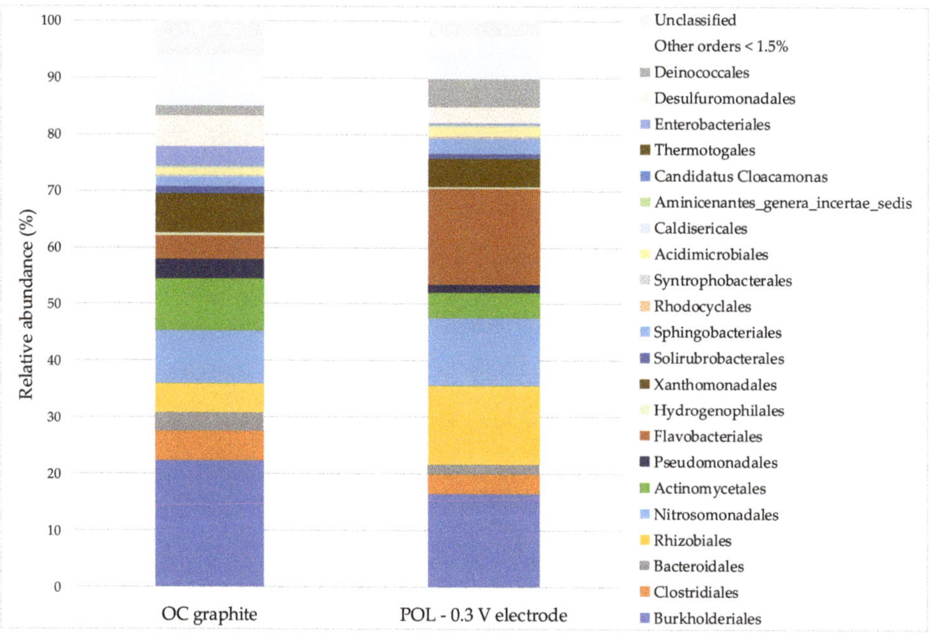

Figure 6. Bacterial community structures at order level of the biofilm developed on graphite of the open circuit system (OC graphite) and on polarized electrode (POL −0.3 V electrode). "Other orders <1.5%" indicated strains with abundance less than 1.5%.

The only genus detected within the *Flavobacteriales* order was *Moheibacter*. *Moheibacter* was isolated in a cathodic biofilm in a BES for the degradation of oxytetracicline [67] and also in underground rocks, and in anaerobic environment [68]. Schauss and colleagues [69], isolated a novel *Moheibacter* sp. whit high content of quinone that could be involved in electron transfer to multi-heme c-type cytochromes that have a key role in External Electron Transfer (EET) to electrodes and minerals [12]. The sharp increase in relative abundances of this genus on the polarized bioelectrode, as well as in other electroactive biocathodic communities, as reported in the literature [70,71], may suggest this genus is involved in the transfer of electrons from an electrode. However, to date, there has been no research that has pointed out the *Moheibacter* is directly implicated in EET. The only genus that was detected within the *Rhizobiales* order was *Nitrobacter*. Bacteria that belong to the *Nitrobacter* genus are known for oxidizing nitrite to nitrate, and have been previously observed in autotrophic biocathodes for nitrate removal [72,73]. The whole *Deinococcales* order was constituted by *Truepera* genus. This genus has been previously described in electroactive cathodic communities also consisting of microorganisms belonging to the genera *Moheibacter* and *Nitrosomonas* [72,74]. Although, an ecological relationship has not yet been defined, the abundance of these genera on the polarized bioelectrode suggests they were at an advantage compared to open circuit system.

Table 2. Relative abundance at order and genus level of the biofilm developed on graphite of the open circuit system (OC graphite) and on polarized electrode (POL −0.3 V electrode).

Order / Genus	OC Graphite (%)	POL −0.3 V Electrode (%)
Burkholderiales	22.4	16.5
Advenella	14.7	8.2
Cupriavidus	2.8	1.6
Polaromonas	1.6	<1.5
Flavobacteriales	4.2	17.0
Moheibacter	4.1	16.7
Nitrosomonadales	9.3	11.9
Nitrosomonas	9.1	11.9
Rhizobiales	5.2	14.0
Nitrobacter	<1.5	9.9
Actinomycetales	9.2	4.4
Rhodococcus	5.0	1.5
Xanthomonadales	6.9	5.0
Stenotrophomonas	3.0	3.1
Dyella	3.3	<1.5
Clostridiales	5.1	3.5
Clostridium XI	1.8	1.8
Desulfuromonadales	5.4	2.9
Geobacter	4.2	1.8
Deinococcales	1.8	4.9
Truepera	1.8	4.9
Bacteroidales	3.2	1.7
Petrimonas	1.7	<1.5
Pseudomonadales	3.3	1.4
Pseudomonas	3.1	<1.5
Sphingobacteriales	1.6	2.9
Enterobacteriales	3.5	<1.5
Escherichia/Shigella	3.5	<1.5
Acidimicrobiales	<1.5	1.8
Other orders <1.5%	8.9	7.5
Others genera <1.5%	24.4	21.9
Unclassified order	6.1	2.7
Unclassified genus	16.1	16.7

4. Conclusions

The acclimatization phase in the MFC allowed for the rapid development of the electroactive biofilm, which was able to couple organic substrate oxidation with bioenergy production. Furthermore, the development of an electroactive biofilm shortened the time for Cr(VI) reduction. The bioelectrode polarized at −0.3 V (versus SHE) reduced dissolved Cr(VI) with larger efficiency (above 90%) and/or less time than the controls. Bioelectrochemical reduction overcame both, purely electrochemical reduction, showing only 35% Cr(VI) removal under neutral pH, and bioreduction, characterized by high overall efficiency and a significantly slower rate. Furthermore, the bioelectrochemical reduction rate was 20% faster than the other mechanisms (biological and electrochemical reduction).

Electroactive biofilm was proven to be essential for the removal of dissolved chromium. Community analyses suggest that *Moheibacter*, *Nitrobacter* and *Truepera* were selectively enriched in the biofilm of the polarized system. Bacteria belonging to the *Flavobacteriales*, *Nitrosomonadales* and *Rhizobiales* orders play a dominant role in electroactive communities enriched in POL −0.3 V. This work was done to demonstrate, on a laboratory scale, that the bioelectrochemical removal of Cr(VI) can occur, even in the absence of organic carbon. The results of this study corroborate the results of previous studies, that reported high efficiency in bioelectrochemical Cr(VI) removal polarized electrodes at −0.3 V. Although, bioelectrochemical Cr(VI) reduction requires further research, especially at the pilot scale with real contaminated groundwater, this approach can be considered the basis for a new and sustainable technology for groundwater remediation.

Supplementary Materials: The following are available online at http://www.mdpi.com/2073-4441/12/2/466/s1, Figure S1: Rarefaction curves of bacterial communities based on OTUS, from the following reactors: t0 Ph. I, t0 Ph. II, POL −0.3 V electrode, POL −0.3 V, OC graphite, OC; Table S1: Species richness and diversity estimates.

Author Contributions: Conceptualization, G.B., A.F.M. and E.S.; data curation, A.E.T. and A.F.; formal analysis, G.B., A.E.T., A.F., A.F.M. and E.S.; funding acquisition, S.S.; investigation, G.B., A.F.M. and E.S.; methodology, G.B., A.F.M. and E.S.; supervision, A.F. and S.S.; visualization, G.B., M.D., A.E.T., A.F.M. and E.S.; writing—original draft, G.B., A.E.T., A.F.M. and E.S.; writing—review and editing, G.B., M.D., A.E.T., A.F.M. and E.S. All authors have read and agree to the published version of the manuscript.

Funding: This research was funded by this work was supported by Fondazione Cariplo in the framework of the project BEvERAGE–BioElEctrochemical RemediAtion of Groundwater plumes (2015-0195).

Conflicts of Interest: The authors declare no conflict of interest.

References

1. Ayyappan, C.S. Microbial leaching of chromium from solidified waste forms—A kinetic study. *J. Ecol. Eng.* **2015**, *16*, 36–42. [CrossRef]

2. Tschersich, C.; Barouki, R.; Uhl, M.; Klánová, J.; Horvat, M.; Alimonti, A.; Sarigiannis, D.; Santonen, T.; Lebret, E.; Schoeters, G. *Scoping Documents for 2018 Deliverable Report D 4.2—Input for Cd and Cr(VI)*; Tenical Report for HBM4EU H2020 Programme; 2018; Available online: https://www.hbm4eu.eu/deliverables/ (accessed on 7 February 2020).

3. Mohan, D.; Rajput, S.; Singh, V.K.; Steele, P.H.; Pittman, C.U. Modeling and evaluation of chromium remediation from water using low cost bio-char, a green adsorbent. *J. Hazard. Mater.* **2011**, *188*, 319–333. [CrossRef]

4. Song, Z.; Williams, C.J.; Edyvean, R.G.J. Sedimentation of tannery wastewater. *Water Res.* **2000**, *34*, 2171–2176. [CrossRef]

5. Madhavi, V.; Vijay, A.; Reddy, B.; Reddy, K.G.; Madhavi, G.; Nagavenkata, T.; Prasad, K.V. An Overview on Research Trends in Remediation of Chromium. *Res. J. Recent Sci. Res.* **2013**, *2*, 71–83.

6. Peng, H.; Guo, J.; Li, B.; Liu, Z.; Tao, C. High-efficient recovery of chromium (VI) with lead sulfate. *J. Taiwan Inst. Chem. Eng.* **2018**, *85*, 149–154. [CrossRef]

7. Peng, H.; Leng, Y.; Cheng, Q.; Shang, Q.; Shu, J.; Guo, J. Efficient Removal of Hexavalent Chromium from Wastewater with Electro-Reduction. *Processes* **2019**, *7*, 41. [CrossRef]

8. Jobby, R.; Jha, P.; Yadav, A.K.; Desai, N. Biosorption and biotransformation of hexavalent chromium [Cr(VI)]: A comprehensive review. *Chemosphere* **2018**, *207*, 255–266. [CrossRef]

9. Daghio, M.; Aulenta, F.; Vaiopoulou, E.; Franzetti, A.; Arends, J.B.A.; Sherry, A.; Suárez-Suárez, A.; Head, I.M.; Bestetti, G.; Rabaey, K. Electrobioremediation of oil spills. *Water Res.* **2017**, *114*, 351–370. [CrossRef]

10. Logan, B.E.; Hamelers, B.; Rozendal, R.; Schröder, U.; Keller, J.; Freguia, S.; Aelterman, P.; Verstraete, W.; Rabaey, K. Microbial fuel cells: Methodology and technology. *Environ. Sci. Technol.* **2006**, *40*, 5181–5192. [CrossRef]

11. Rabaey, K.; Angenent, L.; Schroder, U.; Keller, J. *Bioelectrochemical Systems: From Extracellular Electron Transfer to Biotechnological Application*; IWA Publishing: London, UK, 2009; ISBN 9781843392330.

12. Shi, L.; Dong, H.; Reguera, G.; Beyenal, H.; Lu, A.; Liu, J.; Yu, H.-Q.Q.; Fredrickson, J.K. Extracellular Electron Transfer Mechanisms between Microorganisms and Minerals. *Nat. Rev. Microbiol.* **2016**, *14*, 651–662. [CrossRef]

13. An, Z.; Zhang, H.; Wen, Q.; Chen, Z.; Du, M. Desalination combined with hexavalent chromium reduction in a microbial desalination cell. *Desalination* **2014**, *354*, 181–188. [CrossRef]

14. Li, Z.; Zhang, X.; Lei, L. Electricity production during the treatment of real electroplating wastewater containing Cr6+ using microbial fuel cell. *Process Biochem.* **2008**, *43*, 1352–1358. [CrossRef]

15. Wang, G.; Huang, L.; Zhang, Y. Cathodic reduction of hexavalent chromium [Cr(VI)] coupled with electricity generation in microbial fuel cells. *Biotechnol. Lett.* **2008**, *30*, 1959–1966. [CrossRef] [PubMed]

16. Zhang, B.; Feng, C.; Ni, J.; Zhang, J.; Huang, W. Simultaneous reduction of vanadium (V) and chromium (VI) with enhanced energy recovery based on microbial fuel cell technology. *J. Power Sources* **2012**, *204*, 34–39. [CrossRef]

17. Huang, L.; Chen, J.; Quan, X.; Yang, F. Enhancement of hexavalent chromium reduction and electricity production from a biocathode microbial fuel cell. *Bioprocess Biosyst. Eng.* **2010**, *33*, 937–945. [CrossRef]

18. Tandukar, M.; Huber, S.J.; Onodera, T.; Pavlostathis, S.G. Biological chromium(VI) reduction in the cathode of a microbial fuel cell. *Environ. Sci. Technol.* **2009**, *43*, 8159–8165. [CrossRef]

19. Huang, L.; Chai, X.; Quan, X.; Logan, B.E.; Chen, G. Reductive dechlorination and mineralization of pentachlorophenol in biocathode microbial fuel cells. *Bioresour. Technol.* **2012**, *111*, 167–174. [CrossRef]

20. Xafenias, N.; Zhang, Y.; Banks, C.J. Enhanced performance of hexavalent chromium reducing cathodes in the presence of Shewanella oneidensis MR-1 and lactate. *Environ. Sci. Technol.* **2013**, *47*, 4512–4520. [CrossRef]

21. Thatoi, H.; Das, S.; Mishra, J.; Rath, B.P.; Das, N. Bacterial chromate reductase, a potential enzyme for bioremediation of hexavalent chromium: A review. *J. Environ. Manag.* **2014**, *146*, 383–399. [CrossRef]

22. Lovley, D.R.; Phillips, E.J.P. Reduction of Chromate by Desulfovibrio-Vulgaris and Its c3 Cytochrome. *Appl. Environ. Microbiol.* **1994**, *60*, 726–728. [CrossRef]

23. Chung, J.; Nerenberg, R.; Rittmann, B.E. Bio-reduction of soluble chromate using a hydrogen-based membrane biofilm reactor. *Water Res.* **2006**, *40*, 1634–1642. [CrossRef] [PubMed]

24. Inglett, K.S.; Bae, H.S.; Aldrich, H.C.; Hatfield, K.; Ogram, A.V. *Clostridium chromiireducens* sp. nov., isolated from Cr(VI)-contaminated soil. *Int. J. Syst. Evol. Microbiol.* **2011**, *61*, 2626–2631. [CrossRef] [PubMed]

25. Mclean, J.; Beveridge, T.J. Chromate Reduction by a Pseudomonad Isolated from a Site Contaminated with Chromated Copper Arsenate. *Appl. Environ. Microbiol.* **2001**, *67*, 1076–1084. [CrossRef] [PubMed]

26. McLean, J.S.; Beveridge, T.J.; Phipps, D. Isolation and characterization of a chromium-reducing bacterium from a chromated copper arsenate-contaminated site. *Environ. Microbiol.* **2000**, *2*, 611–619. [CrossRef] [PubMed]

27. Malaviya, P.; Singh, A. Bioremediation of chromium solutions and chromium containing wastewaters. *Crit. Rev. Microbiol.* **2016**, *42*, 607–633. [CrossRef] [PubMed]

28. Wang, H.; Zhang, S.; Wang, J.; Song, Q.; Zhang, W.; He, Q.; Song, J.; Ma, F. Comparison of performance and microbial communities in a bioelectrochemical system for simultaneous denitrification and chromium removal: Effects of pH. *Process Biochem.* **2018**, *73*, 154–161. [CrossRef]

29. Xue, H.; Zhou, P.; Huang, L.; Quan, X.; Yuan, J. Cathodic Cr(VI) reduction by electrochemically active bacteria sensed by fluorescent probe. *Sens. Actuators B Chem.* **2017**, *243*, 303–310. [CrossRef]

30. Pous, N.; Balaguer, M.D.; Colprim, J.; Puig, S. Opportunities for groundwater microbial electro-remediation. *Microb. Biotechnol.* **2018**, *11*, 119–135. [CrossRef]

31. Williams, K.H.; Nevin, K.P.; Franks, A.; Englert, A.; Long, P.E.; Lovley, D.R. Electrode-based approach for monitoring in situ microbial activity during subsurface bioremediation. *Environ. Sci. Technol.* **2010**, *44*, 47–54. [CrossRef]

32. Huang, L.; Wang, Q.; Jiang, L.; Zhou, P.; Quan, X.; Logan, B.E. Adaptively Evolving Bacterial Communities for Complete and Selective Reduction of Cr(VI), Cu(II), and Cd(II) in Biocathode Bioelectrochemical Systems. *Environ. Sci. Technol.* **2015**, *49*, 9914–9924. [CrossRef]

33. Song, T.S.; Jin, Y.; Bao, J.; Kang, D.; Xie, J. Graphene/biofilm composites for enhancement of hexavalent chromium reduction and electricity production in a biocathode microbial fuel cell. *J. Hazard. Mater.* **2016**, *317*, 73–80. [CrossRef] [PubMed]

34. Wu, X.; Zhu, X.; Song, T.; Zhang, L.; Jia, H.; Wei, P. Effect of acclimatization on hexavalent chromium reduction in a biocathode microbial fuel cell. *Bioresour. Technol.* **2015**, *180*, 185–191. [CrossRef] [PubMed]

35. Kellner, K.; Posnicek, T.; Ettenauer, J.; Zuser, K.; Brandl, M. A new, low-cost potentiostat for environmental measurements with an easy-to-use PC interface. *Procedia Eng.* **2015**, *120*, 956–960. [CrossRef]

36. Rowe, A.A.; Bonham, A.J.; White, R.J.; Zimmer, M.P.; Yadgar, R.J.; Hobza, T.M.; Honea, J.W.; Ben-Yaacov, I.; Plaxco, K.W. Cheapstat: An open-source, "do-it-yourself" potentiostat for analytical and educational applications. *PLoS ONE* **2011**, *6*, e23783. [CrossRef]

37. Eaton, A.D.; Franson, M.A.H. *Standard Methods for the Examination of Water and Wastewater*, 21st ed.; American Public Health Association, Ed.; American Public Health Association: Washington, DC, USA, 2005; ISBN 9780875530475.

38. Huber, J.A.; Welch, D.B.M.; Morrison, H.G.; Huse, S.M.; Neal, P.R.; Butterfield, D.A.; Sogin, M.L. Microbial population structures in the deep marine biosphere. *Science* **2007**, *318*, 97–101. [CrossRef]

39. Wang, Y.; Qian, P.Y. Conservative fragments in bacterial 16S rRNA genes and primer design for 16S ribosomal DNA amplicons in metagenomic studies. *PLoS ONE* **2009**, *4*, e7401. [CrossRef]

40. Palma, E.; Daghio, M.; Espinoza Tofalos, A.; Franzetti, A.; Cruz Viggi, C.; Fazi, S.; Petrangeli Papini, M.; Aulenta, F. Anaerobic electrogenic oxidation of toluene in a continuous-flow bioelectrochemical reactor: Process performance, microbial community analysis, and biodegradation pathways. *Environ. Sci. Water Res. Technol.* **2018**, *4*, 2136–2145. [CrossRef]

41. Wang, Q.; Garrity, G.M.; Tiedje, J.M.; Cole, J.R. Naïve Bayesian classifier for rapid assignment of rRNA sequences into the new bacterial taxonomy. *Appl. Environ. Microbiol.* **2007**, *73*, 5261–5267. [CrossRef]

42. Rismani-Yazdi, H.; Carver, S.M.; Christy, A.D.; Tuovinen, O.H. Cathodic limitations in microbial fuel cells: An overview. *J. Power Sources* **2008**, *180*, 683–694. [CrossRef]

43. Lu, L.; Ren, N.; Zhao, X.; Wang, H.; Wu, D.; Xing, D. Hydrogen production, methanogen inhibition and microbial community structures in psychrophilic single-chamber microbial electrolysis cells. *Energy Environ. Sci.* **2011**, 1329–1336. [CrossRef]

44. Wang, Y.-T. Microbial reduction of chromate. In *Environmental Microbe-Metal Interactions*; American Society of Microbiology: Washington, DC, USA, 2000; pp. 225–235.

45. Viti, C.; Marchi, E.; Decorosi, F.; Giovannetti, L. Molecular mechanisms of Cr(VI) resistance in bacteria and fungi. *FEMS Microbiol. Rev.* **2014**, *38*, 633–659. [CrossRef] [PubMed]

46. Huang, L.; Regan, J.M.; Quan, X. Electron transfer mechanisms, new applications, and performance of biocathode microbial fuel cells. *Bioresour. Technol.* **2011**, *102*, 316–323. [CrossRef] [PubMed]

47. Focardi, S.; Pepi, M.; Focardi, S.E. Microbial Reduction of Hexavalent Chromium as a Mechanism of Detoxification and Possible Bioremediation Applications. *Biodegrad. Life Sci.* **2013**, *12*, 321–348.

48. Vendruscolo, F.; da Rocha Ferreira, G.L.; Antoniosi Filho, N.R. Biosorption of hexavalent chromium by microorganisms. *Int. Biodeterior. Biodegrad.* **2017**, *119*, 87–95. [CrossRef]

49. Qu, Y.; Zhang, X.; Xu, J.; Zhang, W.; Guo, Y. Removal of hexavalent chromium from wastewater using magnetotactic bacteria. *Sep. Purif. Technol.* **2014**, *136*, 10–17. [CrossRef]

50. Gangadharan, P.; Nambi, I.M. Hexavalent chromium reduction and energy recovery by using dual-chambered microbial fuel cell. *Water Sci. Technol.* **2015**, *71*, 353–358. [CrossRef]

51. Li, M.; Zhou, S.; Xu, Y.; Liu, Z.; Ma, F.; Zhi, L.; Zhou, X. Simultaneous Cr(VI) reduction and bioelectricity generation in a dual chamber microbial fuel cell. *Chem. Eng. J.* **2018**, *334*, 1621–1629. [CrossRef]

52. Singhvi, P.; Chhabra, M.; Singhvi, P.; Chhabra, M. Simultaneous Chromium Removal and Power Generation Using Algal Biomass in a Dual Chambered Salt Bridge Microbial Fuel Cell. *J. Bioremediat. Biodegrad.* **2013**, *4*, 190.

53. Chao, A. Estimating the Population Size for Capture-Recapture Data with Unequal Catchability. *Biometrics* **1987**, *43*, 783–791. [CrossRef]

54. Sotres, A.; Díaz-Marcos, J.; Guivernau, M.; Illa, J.; Magrí, A.; Prenafeta-Boldú, F.X.; Bonmatí, A.; Viñas, M. Microbial community dynamics in two-chambered microbial fuel cells: Effect of different ion exchange membranes. *J. Chem. Technol. Biotechnol.* **2015**, *90*, 1497–1506. [CrossRef]

55. Barbosa, S.G.; Peixoto, L.; Soares, O.S.G.P.; Pereira, M.F.R.; Heijne, A.T.; Kuntke, P.; Alves, M.M.; Pereira, M.A. Influence of carbon anode properties on performance and microbiome of Microbial Electrolysis Cells operated on urine. *Electrochim. Acta* **2018**, *267*, 122–132. [CrossRef]

56. Read, S.T.; Dutta, P.; Bond, P.L.; Keller, J.; Rabaey, K. Initial development and structure of biofilms on microbial fuel cell anodes. *BMC Microbiol.* **2010**, *10*, 98. [CrossRef] [PubMed]

57. Joicy, A.; Song, Y.-C.; Lee, C.-Y. Electroactive microorganisms enriched from activated sludge remove nitrogen in bioelectrochemical reactor. *J. Environ. Manag.* **2019**, *233*, 249–257. [CrossRef] [PubMed]

58. Saratale, G.D.; Saratale, R.G.; Shahid, M.K.; Zhen, G.; Kumar, G.; Shin, H.S.; Choi, Y.G.; Kim, S.H. A comprehensive overview on electro-active biofilms, role of exo-electrogens and their microbial niches in microbial fuel cells (MFCs). *Chemosphere* **2017**, *178*, 534–547. [CrossRef] [PubMed]

59. Franzetti, A.; Daghio, M.; Parenti, P.; Truppi, T.; Bestetti, G.; Trasatti, S.P.; Cristiani, P. Monod Kinetics Degradation of Low concentration Residual Organics in Membraneless Microbial Fuel Cells. *J. Electrochem. Soc.* **2017**, *164*, H3091–H3096. [CrossRef]

60. Yun, H.; Liang, B.; Kong, D.; Wang, A. Improving biocathode community multifunctionality by polarity inversion for simultaneous bioelectroreduction processes in domestic wastewater. *Chemosphere* **2018**, *194*, 553–561. [CrossRef] [PubMed]

61. Herrmann, M.; Opitz, S.; Harzer, R.; Totsche, K.; Küsel, K. Attached and Suspended Denitrifier Communities in Pristine Limestone Aquifers Harbor High Fractions of Potential Autotrophs Oxidizing Reduced Iron and Sulfur Compounds. *Microb. Ecol.* **2017**, *74*, 264–277. [CrossRef]

62. Fernandez, N.; Sierra-Alvarez, R.; Amils, R.; Field, J.A.; Sanz, J.L. Compared microbiology of granular sludge under autotrophic, mixotrophic and heterotrophic denitrification conditions. *Water Sci. Technol.* **2009**, *59*, 1227–1236. [CrossRef]

63. Im, C.H.; Kim, C.; Song, Y.E.; Oh, S.E.; Jeon, B.H.; Kim, J.R. Electrochemically enhanced microbial CO conversion to volatile fatty acids using neutral red as an electron mediator. *Chemosphere* **2018**, *191*, 166–173. [CrossRef]

64. Paiva, M.C.; Ávila, M.P.; Reis, M.P.; Costa, P.S.; Nardi, R.M.D.; Nascimento, A.M.A. The microbiota and abundance of the class 1 integron-integrase gene in tropical sewage treatment plant influent and activated sludge. *PLoS ONE* **2015**, *10*, e0131532. [CrossRef]

65. Khanongnuch, R.; Di Capua, F.; Lakaniemi, A.-M.; Rene, E.R.; Lens, P.N.L. H$_2$S removal and microbial community composition in an anoxic biotrickling filter under autotrophic and mixotrophic conditions. *J. Hazard. Mater.* **2018**, *367*, 397–406. [CrossRef] [PubMed]

66. Morel, M.A.; Iriarte, A.; Jara, E.; Musto, H.; Castro-Sowinski, S. Revealing the biotechnological potential of Delftia sp. JD2 by a genomic approach. *AIMS Bioeng.* **2016**, *3*, 156–175. [CrossRef]

67. Chen, J.; Yang, Y.; Liu, Y.; Tang, M.; Wang, R.; Tian, Y.; Jia, C. Bacterial community shift and antibiotics resistant genes analysis in response to biodegradation of oxytetracycline in dual graphene modified bioelectrode microbial fuel cell. *Bioresour. Technol.* **2019**, *276*, 236–243. [CrossRef] [PubMed]

68. Mei, J.; Wu, Y.; Qian, F.; Chen, C.; Shen, Y.; Zhao, Y. Methane-Oxidizing Microorganism Properties in Landfills. *Polish J. Environ. Stud.* **2019**, *28*, 3809–3818. [CrossRef]

69. Schauss, T.; Busse, H.J.; Golke, J.; Kämpfer, P.; Glaeser, S.P. Moheibacter stercoris sp. Nov., isolated from an input sample of a biogas plant. *Int. J. Syst. Evol. Microbiol.* **2016**, *66*, 2585–2591. [CrossRef]

70. Liao, C.; Wu, J.; Zhou, L.; Li, T.; Du, Q.; An, J.; Li, N.; Wang, X. Optimal set of electrode potential enhances the toxicity response of biocathode to formaldehyde. *Sci. Total Environ.* **2018**, *644*, 1485–1492. [CrossRef]

71. Sun, J.; Xu, W.; Yang, P.; Li, N.; Yuan, Y.; Zhang, H.; Ning, X.; Zhang, Y.; Chang, K.; Peng, Y.; et al. Enhancing the performance of photo-bioelectrochemical fuel cell using graphene oxide/cobalt/polypyrrole composite modified photo-biocathode in the presence of antibiotic. *Int. J. Hydrogen Energy* **2018**, *44*, 1919–1929. [CrossRef]

72. Huang, L.; Wang, Q.; Quan, X.; Liu, Y.; Chen, G. Bioanodes/biocathodes formed at optimal potentials enhance subsequent pentachlorophenol degradation and power generation from microbial fuel cells. *Bioelectrochemistry* **2013**, *94*, 13–22. [CrossRef]

73. Xiao, Y.; Zheng, Y.; Wu, S.; Yang, Z.H.; Zhao, F. Bacterial Community Structure of Autotrophic Denitrification Biocathode by 454 Pyrosequencing of the 16S rRNA Gene. *Microb. Ecol.* **2015**, *69*, 492–499. [CrossRef]

74. Liao, C.; Wu, J.; Zhou, L.; Li, T.; An, J.; Huang, Z.; Li, N.; Wang, X. Repeated transfer enriches highly active electrotrophic microbial consortia on biocathodes in microbial fuel cells. *Biosens. Bioelectron.* **2018**, *121*, 118–124. [CrossRef]

Article

Comparing the Adsorption Performance of Multiwalled Carbon Nanotubes Oxidized by Varying Degrees for Removal of Low Levels of Copper, Nickel and Chromium(VI) from Aqueous Solutions

Marko Šolić [1], Snežana Maletić [1,*], Marijana Kragulj Isakovski [1], Jasmina Nikić [1], Malcolm Watson [1], Zoltan Kónya [2,3] and Jelena Tričković [1]

[1] Department of Chemistry, Biochemistry and Environmental Protection, Faculty of Sciences, University of Novi Sad, Trg Dositeja Obradovića 3, 21000 Novi Sad, Serbia; marko.solic@dh.uns.ac.rs (M.S.); marijana.kragulj@dh.uns.ac.rs (M.K.I.); jasmina.nikic@dh.uns.ac.rs (J.N.); malcolm.watson@dh.uns.ac.rs (M.W.); jelena.trickovic@dh.uns.ac.rs (J.T.)

[2] Department of Applied and Environmental Chemistry, University of Szeged, Rerrich Béla tér 1, H-6720 Szeged, Hungary; konya@chem.u-szeged.hu

[3] MTA-SZTE Reaction Kinetics and Surface Chemistry Research Group, Rerrich Béla tér 1, H-6720 Szeged, Hungary

* Correspondence: snezana.maletic@dh.uns.ac.rs; Tel.: +381-21-485-2727

Received: 31 January 2020; Accepted: 3 March 2020; Published: 6 March 2020

Abstract: Functionalized multiwalled carbon nanotubes (MWCNTs) have drawn wide attention in recent years as novel materials for the removal of heavy metals from the aquatic media. This paper investigates the effect that the functionalization (oxidation) process duration time (3 h or 6 h) has on the ability of MWCNTs to treat water contaminated with low levels of Cu(II), Ni(II) and Cr(VI) (initial concentrations 0.5–5 mg L^{-1}) and elucidates the adsorption mechanisms involved. Adsorbent characterization showed that the molar ratio of C and O in these materials was slightly lower for the oxMWCNT6h, due to the higher degree of oxidation, but the specific surface areas and mesopore volumes of these materials were very similar, suggesting that prolonging the functionalization duration had an insignificant effect on the physical characteristics of oxidized multiwalled carbon nanotubes (oxMWCNTs). Increasing the Ph of the solutions from Ph 2 to Ph 8 had a large positive impact on the removal of Cu(II) and Ni(II) by oxMWCNT, but reduced the adsorption of Cr(VI). However, the ionic strength of the solutions had far less pronounced effects. Coupled with the results of fitting the kinetics data to the Elowich and Weber–Morris models, we conclude that adsorption of Cu(II) and Ni(II) is largely driven by electrostatic interactions and surface complexation at the interface of the adsorbate/adsorbent system, whereas the slower adsorption of Cr(VI) on the oxMWCNTs investigated is controlled by an additional chemisorption step where Cr(VI) is reduced to Cr(III). Both oxMWCNT3h and oxMWCNT6h have high adsorption affinities for the heavy metals investigated, with adsorption capacities (expressed by the Freundlich coefficient K_F) ranging from 1.24 to 13.2 (mg g^{-1})/(mg l^{-1})^n, highlighting the great potential such adsorbents have in the removal of heavy metals from aqueous solutions.

Keywords: adsorption; heavy metals; carbon nanotubes; adsorption mechanism

---

1. Introduction

Heavy metals, as naturally occurring elements, can originate from both natural and anthropogenic sources. However, the increasing release of these pollutants in more toxic and mobile forms has made anthropogenic sources a worldwide issue [1]. The main anthropogenic sources of heavy metals in the aquatic environment can be attributed to urbanization, industrialization (in particular the wastewater of modern chemical plants) and certain agricultural activities [1,2]. Metal processing and mining contribute to 48% of the total release of contaminants by the European industrial sector [3]. Heavy metals cannot undergo biodegradation processes and accumulate in the environment and living organisms at all levels of the food chain, making this group of pollutants even more noteworthy. A number of health problems may result from human exposure, intake and eventual buildup of non-essential (e.g., Cr(VI) and Ni) and even essential heavy metals (e.g., Cu) [2,4,5].

For this reason, the permissible concentrations of toxic elements in drinking water, surface waters and wastewaters before discharge to recipients, are set to very low levels. For example, the WHO provisional guideline values for Cu, Ni and Cr are 2 mg L^{-1}, 0.07 mg L^{-1} and 0.05 mg L^{-1}, respectively [6]. Emission levels for those metals in the most wastewater streams, according to the best available technologies, have been limited in the range of 0.01 to the 0.5 mg L^{-1} [7,8]. These concentrations are low, and thus, further treatment in order to comply with legislation is a difficult task. Conventional techniques such as chemical precipitation, coagulation and flocculation etc., are not efficient and are not able to remove low concentrations of target pollutants to permissible levels [2,9]. Techniques such as membrane filtration and ion exchange work well at low concentrations and achieve high efficacies, but suffer from disadvantages such as high operational cost and operational problems, etc. [1,10].

Adsorption can be an economically sustainable and viable solution for the removal of heavy metals from aqueous solutions. Adsorption has many advantages, such as simple operation, low cost, good Ph tolerance, and large industrial processing capacity [2,9,11,12]. However, there is still a need to develop and study new adsorbents, as many of the commonly applied adsorbent materials (activated carbons, zeolites, clay minerals, solid by-products of industrial processes and biosorbents) do not show satisfactory performance for the removal of heavy metals present in low concentrations [10]. Nanomaterials and especially carbon nanotubes (CNT) stand out as particularly promising materials to meet the requirements outlined above [4,10,11,13,14]. In order to achieve the desired level of efficiency for heavy metal removal, CNT functionalization has been widely investigated in recent years. The functionalization of CNT has already led to the development of new adsorbents for aqueous heavy metal treatments with state-of-the-art performance [10]. Even so, many review papers have recognized gaps in the current knowledge, whereby the majority of studies conducted concern the adsorption of very high concentrations of the heavy metals investigated, or do not explore the effects of preparation conditions on the functionalized nanomaterials (such as oxidation process duration, etc.) or the combined effects of different parameters on the adsorption process, or do not investigate the potential mechanism and chemistry involved in the heavy metals removal [1,4,10,11,13–16]. Oxidation of CNT is one of the simplest and therefore most economically viable functionalization techniques [10]. Regarding the removal of heavy metals such as Cu(II), Ni(II) and Cr(VI) by CNT, the literature data is very limited, with papers which deal with just one level of CNT oxidation and which investigate adsorption at high concentration levels above 10 mg L^{-1} [17–23].

The aim of this paper was therefore to investigate how the duration of the oxidation process of multiwalled carbon nanotubes (oxMWCNT3h or oxMWCNT6h) effects their use for the treatment of water contaminated with low levels of Cu(II), Ni(II) and Cr(VI), and to give insight into the adsorption mechanism of these metals on the oxidized MWCNTs. In order to achieve this goal, the synthesized adsorbents have been extensively characterized and the adsorption process was characterized by examining the contact time, the heavy metals concentration, and the Ph and ionic strength of the solutions.

2. Materials and Methods

2.1. Materials and Chemicals

Oxidized multiwalled carbon nanotubes (oxMWCNTs) were synthesized (by chemical vapour deposition) at the University of Szeged, according to the method described elsewhere [24]. Functionalization of MWCNTs was also carried out at the aforementioned institution under varying conditions to provide two different adsorbents for this study, following this procedure: (Step 1) 10 g of pristine MWCNTs was mixed with 1 L of cc. HNO_3 for 1 h by means of magnetic stirring, (Step 2) the obtained suspension was oxidized under reflux for 3 h (oxMWCNT3h) or 6 h (oxMWCNT6h), then (Step 3) rinsed with eionized water until neutral Ph was obtained. The oxMWCNTs selected for this study were then used as received. Stock solutions of Cu(II) and Ni(II) (100 mg L^{-1}) were prepared by diluting the appropriate metal standard solutions (1000 mg L^{-1}), while the Cr(VI) stock solution (100 mg L^{-1}) was obtained by dissolving $K_2Cr_2O_7$ in deionized water. The stock solutions were further diluted to the desired metals concentrations. All chemicals used in this research were analytical grade and were purchased from Merck Co. The experiments were performed using ultrapure deionized water (resistivity not less than 17.5 MΩ).

2.2. Adsorbents Characterisation

The specific surface areas, pore sizes, pore volumes and pore-size distributions of the investigated adsorbents were determined from nitrogen adsorption/desorption isotherms at 77 K, acquired by the AutosorbiQ Surface Area Analyzer (Quantachrome Instruments, USA). The specific surface areas were calculated using the multi-point Brunauer–Emmett–Teller (BET) method, while meso and micro pore volumes were obtained by the utilization of the desorption Barrett–Joyner–Halenda (BJH) isotherms and t-test method, respectively. Scanning electron microscopy (SEM) (TM3030, Hitachi High-Technologies, Japan) coupled with energy dispersive spectrometry (EDS) (Bruker Quantax 70 X-ray detector system, Bruker Nano, GmbH Germany) and transmission electron microscopy (TEM) (Philips CM10) were used to examine the materials morphological structures and surface elemental compositions. Identification of the functional groups present on the oxMWCNT3h and oxMWCNT6h was carried out by Fourier transform infrared (FTIR) spectrometry (Thermo-Nicolet Nexus 670 (USA) FTIR spectrometer), in the 4000–400 cm^{-1} range and in a diffuse reflection mode at a resolution of 4 cm^{-1}. The points of zero charge (pHpzc) of the adsorbents were obtained by adjusting the Ph value of 0.1 M $NaNO_3$ solutions in the range from 2 to 6, followed by measuring the Ph change of the $NaNO_3$/oxMWCNTs mixtures after 24 h of contact.

2.3. Adsorption Experiments

All adsorption studies were carried out in batch experiments conducted according to the following general procedure: (Step 1) 5 mg of oxMWCNT3h or oxMWCNT6h was weighed into 40 Ml glass vials, (Step 2) 30 Ml of background solution ($NaNO_3$) was added to each vial, (Step 3) oxMWCNT3h/$NaNO_3$ and oxMWCNT6h/$NaNO_3$ suspensions were subjected to 30 min of sonication (to enhance the dispersion of the adsorbents) (Ultrasons 1 litre bath), followed by stirring at 180 rpm (to pre-equilibrate oxMWCNTs and $NaNO_3$) for 24 h, (Step 4) stock solutions of Cu(II), Ni(II) or Cr(VI) were spiked to achieve desired concentrations of selected heavy metals, (Step 5) the Ph of adsorption systems was adjusted by adding 0.1 and/or 0.01 M NaOH or HNO_3, (Step 6) the adsorption systems were stirred for a certain period of time at 180 rpm and constant temperature of 298 K, (Step 7) the solid and liquid phases were separated by filtering the samples through 0.45 μm cellulose acetate membrane filters, (Step 8) the residual Cu(II), Ni(II) or Cr(VI) concentrations were determined using ICP-MS technique (Agilent Technologies 7700 Series ICP-MS). Method detection limits for the metals investigated were as follows: Cu(II): 0.001 mg L^{-1}, Ni(II): 0.001 mg L^{-1}, and Cr(VI): 0.001 mg L^{-1}.

The following deviations from this general procedure were made in order to investigate a variety of operational parameters: (1) influence of contact time—adsorption systems were stirred for various durations, ranging from 5 min to 24 h for Cu(II) and Ni(II), and in the case of Cr(VI), from 5 min to 168 h (Step 6); (2) influence of initial metal concentration—initial Cu(II), Ni(II) or Cr(VI) concentrations were varied between 0.5 and 5 mg L^{-1} (Step 4); (3) influence of Ph—initial sample Ph was adjusted to different values, ranging from 2 to 8 (Step 5); (4) influence of ionic strength—the $NaNO_3$ concentration in the background solutions was varied between 0.1 and 1 M (Step 2). Unless otherwise indicated, operational parameters were: (1) contact time—24 h (Cu(II) and Ni(II)) and 96 h (Cr(VI)), (2) 1 mg L^{-1}

initial metal concentration, (3) initial Ph = 5 (Cu(II) and Ni(II)) and Ph = 2.5 (Cr(VI)), (4) 0.1 M NaNO$_3$ ionic strength. The decision to investigate the effect of Ph and ionic strength in this work was made on the basis of the importance that the scientific literature assigns to these parameters.

In order to establish that the observed removals of heavy metals were a consequence of their adsorption on oxMWCNT3h and oxMWCNT6h, and not the result of some other processes (e.g., adsorption on glass vials walls, precipitation, etc.), triplicate blank experiments without the addition of the adsorbents were performed, according to the procedures given above. The levels of Cu(II), Ni(II) or Cr(VI) detected in these runs were used in all calculations as initial adsorbate concentrations. Experimental uncertainty, including instrumental errors, were determined using triplicates of the batch experiments and control samples in each adsorption series, with relative standard deviations falling mostly within ±5% of the reported values.

The adsorption capacities of oxMWCNT3h and oxMWCNT6h for the heavy metals investigated were calculated using mass balance Equation (1). Removal efficiencies of Cu(II), Ni(II) or Cr(VI) were determined by Equation (2).

$$q_t(or\ q_e) = \frac{(C_0 - C_t(or\ C_e))}{m}V \tag{1}$$

$$RE_t(or\ RE_e) = \frac{(C_0 - C_t(or\ C_e))}{C_0} \times 100 \tag{2}$$

where: q_t and q_e represent adsorption capacity per gram dry weight of the adsorbents at a specific time and at the state of equilibrium (mg g^{-1}); C_0, C_t and C_e are the initial, specific time and equilibrium adsorbate concentrations in the liquid phase (mg L^{-1}); V is the volume of the liquid phase (L); m is the dry weight of the adsorbent (g); RE_t and RE_e are the specific time and equilibrium adsorbate removal efficiencies (%).

2.4. Adsorption Kinetics and Isotherms Modelling

2.4.1. Adsorption Kinetics

In order to identify the adsorbate uptake rate, the rate-controlling step and to obtain insight into the possible mechanisms/reaction pathways of the investigated adsorption processes, four kinetic models (Lagergren pseudo-first order, pseudo-second order, Elovich and Weber–Morris (intra-particle diffusion)) were used to fit the experimental data.

The non-linear equations and parameters of the kinetic models tested, as well as information concerning their relationship to the potentially rate limiting adsorption steps, are provided in Table 1.

Table 1. Adsorption kinetics models.

Model	Model Equation	Model Parameters
Lagergren pseudo-first order [a]	$q_t = q_e\left(1 - e^{-k_1 t}\right)$	k_1—Lagergren pseudo-first order rate constant (h^{-1})
Pseudo-second order [b]	$q_t = \frac{q_e^2 k_2 t}{1 + k_2 t q_e}$ $h = k_2 q_e^2$	k_2—pseudo-second order rate constant (g mg^{-1} h^{-1}) h—initial adsorption rate (mg g^{-1} h^{-1})
Elovich [c]	$q_t = \frac{1}{\beta}\ln(1 + \alpha\beta t)$	α—initial adsorption rate (g mg^{-1} h^{-2}) β—extent of surface coverage and the activation energy for chemisorption (g mg^{-1})
Weber–Morris (intra-particle diffusion) [d]	$q_t = k_3 t^{0.5}$	k_3—intra-particle diffusion rate constant (mg g min$^{0.5}$)

Note: Potential steps involved in heavy metal adsorption by porous adsorbents: (1) transport from bulk liquid phase to the external surface of the adsorbent; (2) passage through the liquid film attached to the solid surface; (3) interactions with the adsorbent surface (limiting step in reaction-based models [a], [b] and [c]; (4) diffusion into adsorbent internal sites (pores and interstitial channels) (limiting step in the diffusion-based [d] model) [25].

2.4.2. Adsorption Isotherms

In order to obtain data concerning the maximum adsorption capacity of the adsorbents, as well as to get a better understanding of the adsorption mechanisms, the Freundlich, Langmuir and Dubinin–Radushkevic (D–R) isotherm models were used to analyze the experimental equilibrium data.

The non-linear mathematical expressions, parameters and the most significant assumptions of the three applied models, along with the equations for the separation factor (also called the equilibrium parameter) and adsorption activation energy, which are essential features of the Langmuir and D–R isotherms, are given in Table 2.

Table 2. Adsorption isotherm models.

Model	Model Equation	Model Parameters
Freundlich [a]	$q_e = K_F C_e{}^n$	K_F—Freundlich constant indicating the adsorption capacity ($(mg\ g^{-1})\ (L\ mg^{-1})^{1/n}$) n—Freundlich exponent related to the energy distribution of adsorption sites (adsorption intensity) (dimensionless)
Langmuir [b]	$q_e = \dfrac{q_m K_L C_e}{1 + K_L C_e}$ $R_L = \dfrac{1}{1 + K_L C_0}$	q_m—maximum adsorption capacity ($mg\ g^{-1}$) K_L—Langmuir constant representing the energy of adsorption process ($L\ mg^{-1}$) R_L—Langmuir separation factor associated with adsorption favourability (dimensionless)
D–R [c] [*]	$q_e = q_d{}^{-K_D RT\ ln(1+1/C_e)}$ $E_a = \dfrac{1}{\sqrt{2K_D}}$	q_d—maximum adsorption capacity ($mg\ g^{-1}$) K_D—D–R constant corresponding to the mean free energy of adsorption ($mol2\ K^{-1}\ J^{-2}$) E_a—activation energy ($kJ\ mol^{-1}$)

[a] An empirical equation describing multilayer adsorption onto energy-heterogeneous surfaces with interactions between adsorbed species. The stronger binding sites are occupied first and the binding strength decreases as site occupation increases. [b] A theoretical equation describing monolayer adsorption onto energy-homogenous surfaces with no lateral interaction and steric hindrance between the adsorbed molecules, even on adjacent sites. All binding sites are characterized by the same energy, which makes the binding strength identical throughout the adsorption process. [c] A semi-empirical equation describing adsorption with Gaussian energy distribution onto heterogeneous surfaces. It is usually applied to differentiate between physical and chemical adsorption of metal ions [26]. (*) R is the universal gas constant ($8.314\ J\ mol^{-1}\ K^{-1}$).

3. Results and Discussion

3.1. Adsorbents Characterization

The TEM images of oxMWCNT3h and oxMWCNT6h are shown in Figure 1. The tubes are long and curved forming entangled oxMWCNTs networks. Note that the tubes have open ends as a result of the chemical functionalization (oxidation with cc. HNO_3) [17]. The opening up of the tubes has led to defects on the sidewalls of the nanotubes and has shortened their length. The inner tube diameters of oxMWCNT3h and oxMWCNT6h, ranged from 7–12 and 9–18 nm respectively, while the outer tube diameters of these materials were 15–24 and 7–32 nm [27].

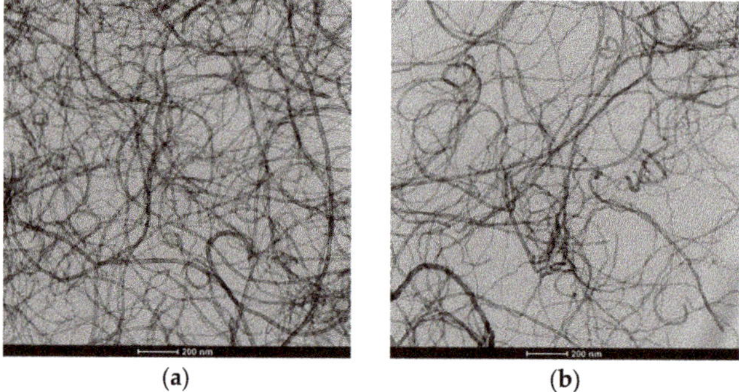

(a)   (b)

Figure 1. TEM images of: (**a**) oxMWCNT3h and (**b**) oxMWCNT6h.

Scanning electron microscopy was also used to investigate the surface morphology of oxMWCNT3h and oxMWCNT6h (Figure S1). It is evident that the surface of oxMWCNTs mainly consists of aggregated nanotubes. EDS analysis of oxMWCNT3h and oxMWCNT6h confirms that C and O are the dominant elements on the surface of these materials. The molar ratios of C and O in these materials were lower for oxMWCNT6h (11.32:1 compared to 9.32:1), as a consequence of the higher rate of oxidation. Traces of N originates from the HNO_3 which was used for functionalization of oxMWCNTs (Table S1).

The FTIR spectra of oxMWCNT3h and oxMWCNT6h are presented in Figure 2. Both spectra exhibit broad peaks at 3420 cm^{-1} which can be assigned to -OH stretching vibration of carboxylic and phenolic groups (-COOH and -COH). Absorption peaks at 2973 and 2923 cm^{-1} relate to the asymmetric and symmetric stretching vibrations of C–H originating from the surface of the tubes or from the sidewalls [20,28]. The peaks at 1630 and 1633 cm^{-1} can be attributed to bending vibration of –OH groups from physisorbed water molecules. The peaks at 1574, 1386, and 1396 cm^{-1}, observed in FTIR spectra of oxMWCNT3h, are associated with asymmetric and symmetric vibration of –COO groups [17,28]. Absorption peaks at 1046, 1086, and 1163 cm^{-1} in FTIR spectra of oxMWCNT3hare are associated with deformation vibrations of the –OH in alcohol, phenolic and carboxyl groups [20,21,28]. Generally, these oxygen-containing functional groups provide numerous adsorption sites and thus increase the adsorption capacity of the MWCNTs for metal ions [17,20].

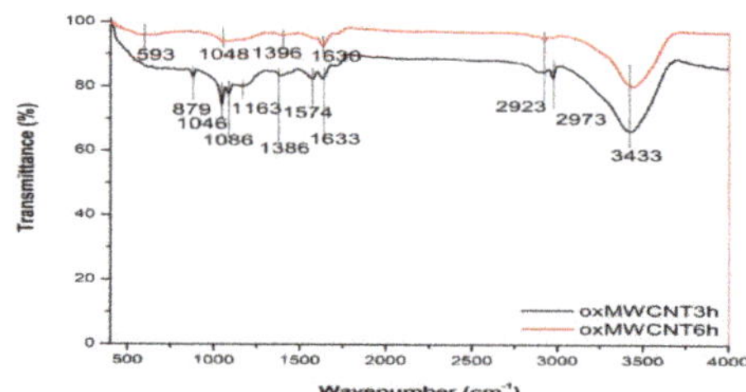

Figure 2. FTIR spectra of oxMWCNT3h and oxMWCNT6h.

The textural characteristics of oxMWCNT3h and oxMWCNT6h, including specific surface area, mesopore and micropore volumes and average pore size are shown in Table 3.

The specific surface areas of oxMWCNT3h and oxMWCNT6h, as well as the mesopore volume of these materials, were very similar, implying that the prolonged mixing time during functionalization has an insignificant effect on the physical characteristics of oxMWCNTs (Table 1). Moreover, the specific surface area of these materials is significantly higher than the specific surface area of MWCNTs which functionalized differently, using H_2O_2 and HNO_3 (in a ratio of 1:3 (v/v)) (196 m^2 g^{-1}) [17]. The point of zero charges of oxMWCNT3h and oxMWCNT6h are 3.2 and 3.4, respectively (Figure S2).

Table 3. Textural properties of oxMWCNT3h and oxMWCNT6h.

		oxMWCNT3h	**oxMWCNT6h**
Specific surface area	$(m^2\,g^{-1})$	277	273
Mesopore volume (2–50 nm)	$(cm^3\,g^{-1})$	1.95	1.65
Micropore volume (<2 nm)	$(cm^3\,g^{-1})$	8×10^{-3}	1×10^{-3}
Average pore size	(nm)	14.2	12.2

3.2. Influence of Contact Time and Investigation of Adsorption Kinetics

3.2.1. Influence of Contact Time

The influence of contact time on the removal of Cu(II), Ni(II) and Cr(VI) by oxMWCNT3h and oxMWCNT6h, presented as q_t vs. t, is shown in Figure 3.

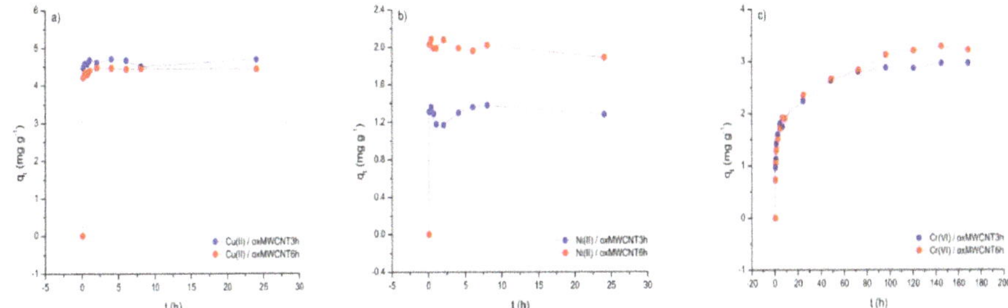

Figure 3. Adsorption of: (**a**) Cu(II), (**b**) Ni(II) and (**c**) Cr(VI) on oxMWCNT3h and oxMWCNT6h as a function of contact time (m = 5 mg, V = 30 mL (0.1 M NaNO$_3$), C_0 = 1 mg L^{-1}, pH = 5 ± 0.1 (Cu(II) and Ni(II)) and 2.5 ± 0.1 (Cr(VI)), t = 5 min-24 h (Cu(II) and Ni(II)) and 5 min-168 h (Cr(VI)), agitation speed = 180 rpm, T = 298 ± 2 K).

As can be seen in Figure 3, in the case of both investigated adsorbents, equilibrium for Cu(II) and Ni(II), was reached very quickly, in less than 20 min, indicating that the adsorption of these metals is predominantly controlled by the interactions taking place on the adsorbate/adsorbent surface. Throughout the entire studied duration, q_t (and RE_t) values remained very stable, averaging 4.60 mg g^{-1} (85.1%), 4.39 mg g^{-1} (93.0%), 1.29 mg g^{-1} (25.2%) and 2.00 mg g^{-1} (39.5%) for Cu(II) and Ni(II), respectively. Unlike in the case of Cu(II), the difference in the affinity of the studied oxMWCNTs for Ni(II) proved evident already in this type of adsorption experiments. Similar results have been obtained by other authors [23,25,29].

In contrast, the overall Cr(VI) adsorption rate for the two examined oxMWCNTs was much lower, with approximately 96 h required to reach equilibrium. Over the monitored period of time, the rate of Cr(VI) removal changed—during the first 8 h it was very high, and then slowly started to decline as the systems gradually approached a state of equilibrium. This behavior immediately suggests that Cr(VI) adsorption kinetics are controlled by more than one process of different physical origins (e.g., film diffusion, surface interactions and intra-particle diffusion), or that they are limited by the action of a single, but highly complex phenomenon (e.g., the presence of surface interactions or diffusion within pores of different dimensions) [30]. Clearly, the increased contact time was followed by an increase in q_t (and RE_t) values, resulting in equilibrium averages (q_e and RE_e) of 2.92 mg g^{-1} (47.8%) and 3.21 mg g^{-1} (58.5%) for oxMWCNT3h and oxMWCNT6h, respectively.

It is also important to note that the duration of the MWCNTs oxidation treatment did not affect the rate at which the equilibrium was reached in any of adsorption systems investigated (oxMWCNT3h = oxMWCNT3h; Cu(II) = Ni(II) > Cr(VI)). Based on the results obtained by these experiments, in order to ensure the feasibility of subsequent experiments, the time given to reach equilibrium in the studies considering the influence of other selected operational parameters for Cu(II)/Ni(II) and Cr(VI) was set to be 24 and 96 h.

3.2.2. Adsorption Kinetics Modelling

Due to very rapid achievement of equilibrium, it was not possible to properly fit the Cu(II) and Ni(II) kinetics data to the kinetics models investigated. Further investigation of adsorption kinetics, utilizing the Lagergren pseudo-first order, pseudo-second order, Elovich and intra-particle diffusion models, was therefore only conducted in the case of Cr(VI).

Based on the R^2 values obtained by non-linear regression of the selected functions (Figure 4a,b), the best fit model was chosen. The calculated parameters of the corresponding equations are listed in Table 4. All the experimental data are in very good compliance with the Elovich model ($R^2 > 0.97$), which is generally valid for systems that are characterized by active chemical adsorption (without desorption of products), a process which includes the occurrence of valence forces, forming through sharing or exchanging of electrons between the adsorbate and the energetically heterogeneous surface of the adsorbent. The suitability of this model, additionally confirmed by the low χ^2 values (0.16 and 0.14; χ^2 determined by means of an χ^2 distribution table), also suggests that the observed decrease in Cr(VI) adsorption rate overtime on both oxMWCNTs is mainly due to the presence of different, complex surface interactions, the most important of which, in this respect, probably relates to very slow Cr(VI)/oxMWCNTs redox reactions (see Section 3.4.) [31–33]. When it comes to Elovich coefficients, it is noticeable that the β values are similar for the two applied adsorbents, whereas the α is more than twice as high for oxMWCNT3h (supported by comparable behavior in the pseudo-second order h values), therefore indicating a faster initial uptake for oxMWCNT3h.

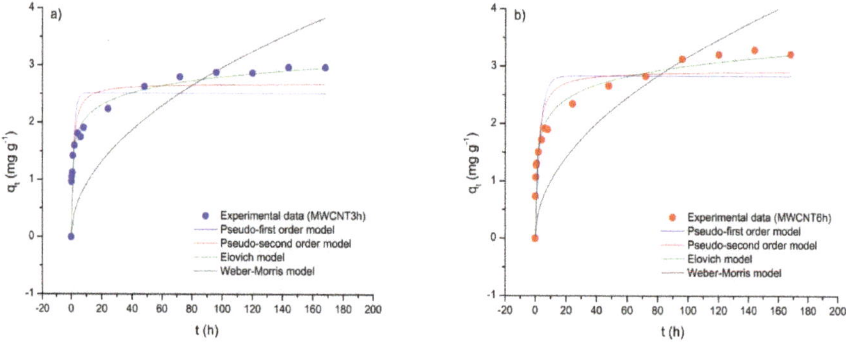

Figure 4. Non-linear regressions of kinetics data, obtained using Lagergren pseudo-first order, pseudo-second order, Elovich and intra-particle diffusion models for Cr(VI) adsorption on (**a**) oxMWCNT3h and (**b**) oxMWCNT6h.

It is important to note that in most of the scientific literature, the intra-particle diffusion model is linearly fitted [34]. According to this approach, intra-particle diffusion plays a significant role in controlling the kinetics only if q_t vs. $t^{0.5}$ produces a straight line passing through the origin of the plot (a zero intercept). Put differently, the deviation from linearity indicates that the adsorption rate is limited by some other process or processes. Furthermore, many authors state that the occurrence of multilinearity in Weber–Morris plots confirms the involvement of intra-particle diffusion in investigated adsorption mechanism, but again, not it the context of the rate-limiting step [35].

Table 4. Parameters of Lagergren pseudo-first order, pseudo-second order, Elovich and intra-particle diffusion models for the adsorption of Cr(VI) on oxMWCNT3h and oxMWCNT6h.

		oxFMWCNT3h	oxFMWCNT6h
Lagergren pseudo-first order	R^2	0.711	0.741
	k_1 (h^{-1})	0.697	0.322
	χ^2	2.051	3.750
	R^2	0.833	0.851
Pseudo-second order	R^2	0.833	0.851
	k_2 (g mg^{-1} h^{-1})	0.329	0.204
	h (g mg^{-1} h^{-1})	2.384	1.756
	χ^2	1.330	1.493
Elovich	R^2	0.981	0.984
	α (g mg^{-1} h^{-2})	30.600	14.562
	β (mg g^{-1} h^{-1})	3.295	2.760
	χ^2	0.160	0.142
Weber–Morris (non-linear)	$R_1{}^2$	0.066	0.342
	k_i (mg g^{-1} h$^{-0.5}$)	0.296	0.317
	χ^2	6.889	6.216
Weber–Morris (linear)	$R_1{}^2$	0.902	0.926
	k_i (mg g^{-1} h$^{-0.5}$)	0.155	0.183
	C_i (mg g^{-1})	1.270	1.195
	χ^2	13.003	11.490

The results obtained using this approach are shown in Figure 5a,b, with the corresponding parameters given in Table 4. As can be seen, both plots are multilinear (the red dashed lines) with the linear fit across the whole range of values yielding an intercept greater than zero (in Table 2 designated C_i). Therefore, as stated in the literature, it can be concluded that intra-particle diffusion, even if not the rate dominant factor, may be involved in the adsorption of Cr(VI) on oxMWCNT3h and oxMWCNT6h. Given the results shown in the Elowich and Weber–Morris models, these adsorption processes (in terms of rate and mechanism) are most likely controlled by complex surface interactions [20,36,37].

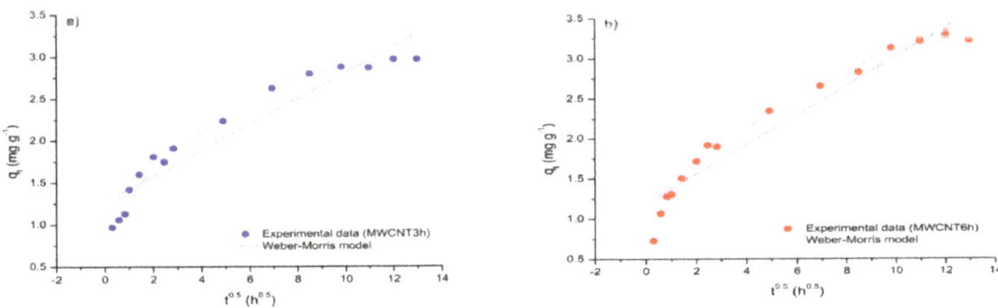

Figure 5. Linear regressions of experimental kinetics data, obtained using the intra-particle diffusion model for Cr(VI) adsorption on: (**a**) oxMWCNT3h and (**b**) oxMWCNT6h.

3.3. Modelling the Adsorption Isotherms (Influence of Initial Metal Concentration)

In order to describe the adsorption of Cu(II), Ni(II) and Cr(VI) on the oxMWCNTs investigated more precisely, the experimental data was analyzed using the Freundlich, Langmuir and D–R equations. The non-linear fitting curves resulting from these models are shown in Figure 6, while the parameters of these functions are given in Table 5.

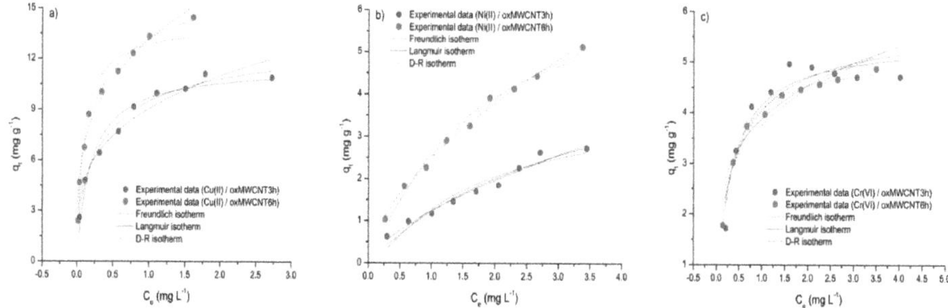

Figure 6. Non-linear regressions of experimental isotherms, obtained using Freundlcih, Langmuir and D–R models for: (**a**) Cu(II), (**b**) Ni(II) and (**c**) Cr(VI) adsorption on oxMWCNT3h and oxMWCNT6h ($m = 5$ mg, $V = 30$ mL (0.1 M NaNO$_3$), $C_0 = 0.5$–5 mg L^{-1}, pH = 5 ± 0.1 (Cu(II) and Ni(II)) and 2.5 ± 0.1 (Cr(VI)), $t = 24$ h (Cu(II) and Ni(II)) and 96 h (Cr(VI)), agitation speed = 180 rpm, T = 298 ± 2 K).

Table 5. Parameters of Freundlich, Langmuir and D–R models for the adsorption of (**a**) Cu(II), (**b**) Ni(II) and (**c**) Cr(VI) on oxMWCNT3h and oxMWCNT6h.

		oxFMWCNT3h			oxFMWCNT6h		
		Cu(II)	Ni(II)	Cr(VI)	Cu(II)	Ni(II)	Cr(VI)
Freundlich	R^2	0.951	0.970	0.706	0.976	0.993	0.920
	K_F (*)	9.066	1.242	3.892	13.165	2.506	3.777
	n	0.281	0.662	0.225	0.286	0.598	0.244
	χ^2	0.474	0.062	0.858	0.695	0.031	0.217
Langmuir	R^2	0.946	0.954	0.920	0.946	0.988	0.994
	K_L (l mg^{-1})	5.129	0.273	3.331	10.459	0.428	3.996
	q_{max} (mg g^{-1})	11.567	5.691	5.422	14.086	8.455	5.100
	R_L	0.867–0.067	0.927–0.516	0.600–0.070	0.905–0.056	0.900–0.409	0.658–0.067
	χ^2	0.735	0.145	0.209	1.124	0.075	0.010
D–R	R^2	0.985	0.922	0.863	0.991	0.975	0.986
	q_d (mg g^{-1})	13.017	4.014	5.764	17.238	7.035	5.509
	K_d (mol^2 K^{-1} J^{-2})	1.846×10^{-4}	6.683×10^{-4}	2.051×10^{-4}	1.555×10^{-4}	5.765×10^{-4}	1.920×10^{-4}
	E_a (kJ mol^{-1})	52.050	27.353	49.376	56.702	29.450	51.027
	χ^2	0.112	0.252	0.369	0.212	0.157	0.037

(*) (mg g^{-1})/(mg l^{-1})^n.

High R^2 (> 0.91) and low χ^2 (< 1.12) values were obtained for all isotherms. The exception is the Freundlich and D–R isotherms of Cr(VI)/oxMWCNT3h, but even here the relatively low R^2 values (0.71 and 0.86) have corresponding acceptable χ^2 (0.86 and 0.37) values, suggesting that each of the three applied models is suitable for describing the adsorption of all the metals investigated on both oxMWCNT3h and oxMWCNT6h. The differences in the goodness of fit for Freundlich and Langmuir models are insignificant, so it is not possible to draw strong conclusions relating and number of adsorbate layers forming during the process.

Values of the Freundlich exponent n are less than 1 in each case, suggesting that all the processes can be characterized as favorable, with the adsorption of Cu(II) and Cr(VI) being good ($0.1 < n < 0.5$) and Ni(II) moderate ($0.5 < n < 1$) (Cu(II) ~ Cr(VI) > Ni(II)). This observation can be further substantiated by the R_L factors, which lie within the range 0 to 1 (adsorption is irreversible, favorable, linear or unfavorable, when $R_L = 0$, $0 < R_L < 1$, $R_L = 1$, $R_L > 1$, respectively). The generally high and significantly different K_F and K_L values obtained indicate the presence of selective adsorbents which have high affinities for the adsorption of the metals investigated.

Depending on how it is expressed, the adsorption capacity of oxMWCNT3h and oxMWCNT6h towards the selected metals exhibits the following trend: K_F (relative)—Cu(II) > Cr(VI) > Ni(II), q_m—Cu(II) > Ni(II) > Cr(VI), q_d—Cu(II) > Cr(VI) > Ni(II) (oxMWCNT3h) and Cu(II) > Ni(II) > Cr(VI) (oxMWCNT6h). Furthermore, the Cu(II) and Ni(II) removals were more pronounced for oxMWCNT6h, while the Cr(VI) uptake was slightly higher in the case of oxMWCNT3h (K_F, q_m, as well as q_d for the adsorption of Cr(VI) on the applied adsorbents differs much less compared to the two other metals). These differences in the removal capacities of oxMWCNT3h and oxMWCNT6h (observed in the case of Cu(II) and Ni(II), and to a much lesser extent for Cr(VI)) are most likely due to the influence of the functionalization process length on the surface properties of the nanotubes (e.g., an increase in the quantity of various oxygen-containing groups with longer functionalization time; see Table S1), making them more suitable for the uptake of the divalent metals studied. Of the trends mentioned above, the experimental observations (q_e values) are most consistent with the values obtained by the Freundlich isotherm, so the assumptions behind this model can potentially be considered the most relevant for this study (see Table 2) [37,38].

Adsorption energies E_a, calculated using the D-R model, are greater than 16 kJ mol^{-1}, which implies that the formation of chemical bonds, as opposed to ion-exchange or non-specific physical interactions, dominantly controls all considered systems (adsorption is physical, ion-exchange or chemical type when 1 kJ mol^{-1} < E_a < 8 kJ mol^{-1}, 8 kJ mol^{-1} < E_a < 16 kJ mol^{-1}, E_a > 16 kJ mol^{-1}, respectively). This finding is consistent with the conclusions reached through non-linear modelling of the kinetics data [39].

A comprehensive overview of the capacity of other available adsorbents is given in the following papers [5,40,41]. As can be seen, oxMWCNT3h and oxMWCNT6h are, in most cases, characterized by a similar or lower adsorption capacity for Cu(II), Ni(II) and Cr(VI) compared to other materials. Considering the non-uniformity of the experimental conditions applied in different studies (e.g., C_0 range, initial pH value, m/V ratio, etc.), as well as the specificity of those under which this research was conducted (very low and narrow C_0 range and low m/V ratio) it is clear that the direct comparison of, for example, the q_{max} values obtained herein with the results of other authors, though possible, provides an unrealistic picture when it comes to efficiency of investigated oxMWCNTs.

3.4. Influence of pH

The solution pH acidity is the one matrix property with the highest potential to affect the adsorption of heavy metals. The removal of Cu(II), Ni(II) and Cr(VI) by oxMWCNT3h and oxMWCNT6h, as a function of pH, is illustrated in Figure 7.

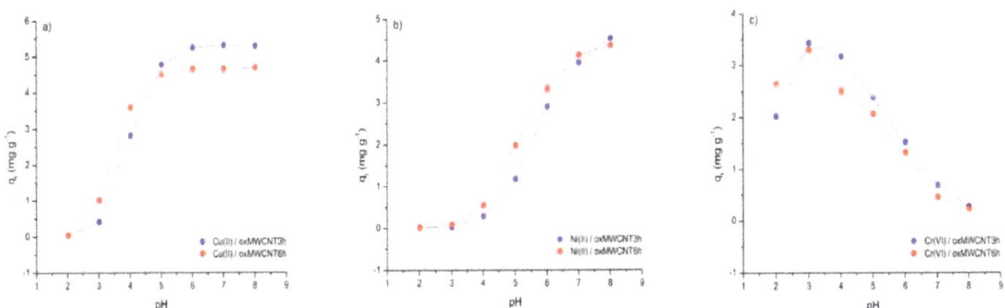

Figure 7. Adsorption of (**a**) Cu(II), (**b**) Ni(II) and (**c**) Cr(VI) on oxMWCNT3h and oxMWCNT6h as a function of pH (m = 5 mg, V = 30 mL (0.1 M NaNO$_3$), C_0 = 1 mg L^{-1}, pH = 2–8 ± 0.1, t = 24 h (Cu(II) and Ni(II)) and 96 h (Cr(VI)), agitation speed = 180 rpm, T = 298 ± 2 K).

As can be seen, pH plays an important role in the behavior of all studied adsorption systems. More specifically, Cu(II) uptake (q_e and RE_e) sharply increases as initial pH goes from 2 to 6 and then, as the liquid phase becomes more basic, maintains a constant high level (0.04 to 5.28 mg g^{-1} (0.70 to 99.4%) and 0.04 to 4.68 mg g^{-1} (0.92 to 99.3%)). oxMWCNTs capacity for Ni(II) rises throughout the entire pH range investigated (0.02 to 4.53 mg g^{-1} (0.40% to 82.2%), whereas the Cr(VI) systems demonstrated the best removals at pH = 3 (3.43 (59.9%) and 3.30 (64.1%)).

The influence of pH on adsorption can be explained by considering the effect this parameter has on the surface properties of the oxMWCNTs and the hydrolysis capacity of the selected metals. Depending on the solution pH, oxygen-containing functional groups (e.g., -OH and -COOH), the presence of which was confirmed in the FTIR spectra, undergo protonation or deprotonation reactions, making the adsorbents net surface charge positive or negative depending upon where the solution pH lies in relation to the adsorbent pHpzc (Equations (3) and (4)).

$$Cx - OH + H^+ \leftrightarrow Cx - OH_2^+ \tag{3}$$

$$Cx - OH \leftrightarrow C_x - O^- + H^+ \tag{4}$$

where: $Cx–OH$, $Cx–OH_2^+$ and $Cx–O^-$ are neutral, protonated and deprotonated sites on the oxMWCNTs surfaces (Cx is the adsorbent carbon surface, while –OH represents all the oxygen-containing functional groups).

Whether, and to what extent, certain heavy metals will interact with such adsorption centers, largely depends on the form/forms in which the metals are present in solution. Therefore, the Visual MINTEQ (version 3.1) software was used to predict the relative proportion of Cu(II), Ni(II) and Cr(VI) species present, with the results shown in Figure 8.

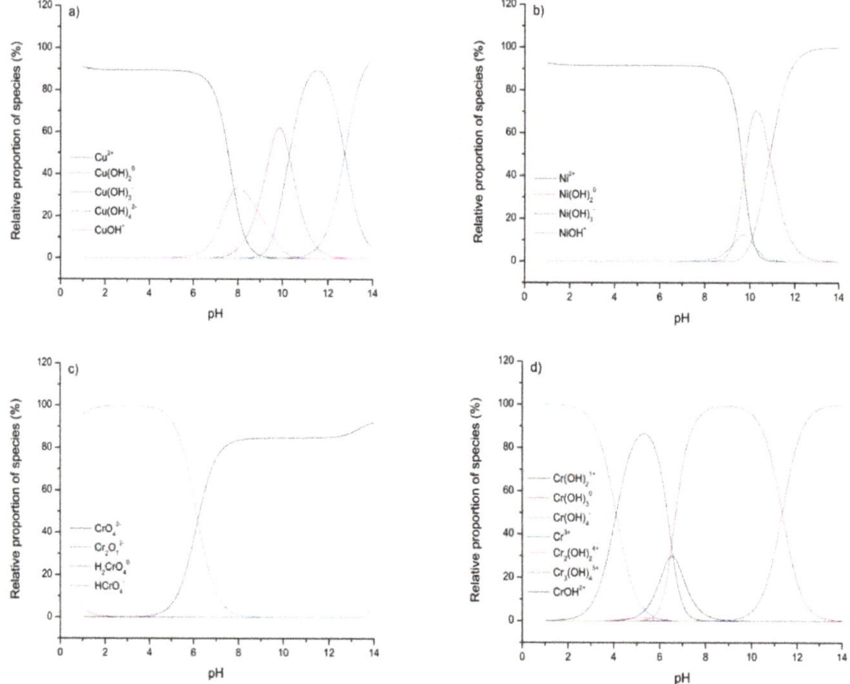

Figure 8. Relative proportion of: (**a**) Cu(II), (**b**) Ni(II), (**c**) Cr(VI) and (**d**) Cr(III) species as a function of pH ($C_0 = 1$ mg L^{-1}, ionic strength = 0.1 M NaNO$_3$, T = 298 K).

According to the plots in Figure 8a,b, at different pH values Cu(II) and Ni(II) exists as M^{2+}, $M(OH)^+$, $M(OH)_2^0$, $M(OH)_3^-$ and/or $M(OH)_4^{2-}$ (Cu(II) only) (M stands for metal). Hence, the low removal of Cu(II) and Ni(II) in strongly acidic media (pH < 4) can be partly attributed to the existence of competition between H^+ and M^{2+} aqua cations ($[(M(H_2O_6)]^{2+}$), the dominant form of these metals under given conditions, for the same oxygen-containing groups. The uptake of highly mobile H^+ ions is clearly more preferential than M^{2+} complex uptake, so most of oxMWCNTs surface sites become occupied and also positively charged (Equation (3)). This therefore hinders the adsorption of Cu(II) and Ni(II) due to the unavailability of sites and electrostatic repulsion. In support of this, the pHpzc of the applied adsorbents was found to be very similar, ~3.2 and ~3.4, indicating that at low pH values oxMWCNT3h and oxMWCNT6h particles exhibit a net positive charge, which makes them incompatible to react with (adsorb) the divalent metals studied. The generally poor, but still evident removals of Cu(II) and Ni(II) in solutions of pH < 4 can be mainly attributed to their exchange with H^+ ions present on the oxMWCNTs surface (Equation (5)).

$$2Cx - OH_2^+ + M^{2+} \leftrightarrow [(Cx - OH)_2M]^+ + 2H^+ \tag{5}$$

When the pH goes up (H^+ concentration decreases), H^+/M^{2+} competition gradually begins to decline and certain acidic functional groups become more dissociated (Equation (4)), thus changing the net surface charge to negative (pH > pHpzc). In an environment like this, cationic metal ions, besides being electrostatically attracted by polarized, but not deprotonated centers (e.g., hydroxyl groups, pKa 9.5 to 13), easily form metal-ligand complexes with the now available, ionized nucleophilic sites, where the CNT oxygens act as electron-donating atoms (e.g., carboxyl (carboxylate) groups, pKa 1.7 to 4.7) (Equation (6) and only in the case of Cu(II) Equation (7)).

$$2C_x - CO^- + M^{2+} \leftrightarrow [(C_x - CO)_2M]^0 \tag{6}$$

$$C_x - CO^- + M(OH)^+ \leftrightarrow [C_x - COM(OH)]^0 \tag{7}$$

Note that beyond pH 7 Cu(II) begins to precipitate as relatively weakly soluble $Cu(OH)_2^0$ due to the more intense hydrolyzation. In contrast, Ni(II) remains stable over the entire tested pH range. Therefore, the removal of Cu(II) in more alkaline (pH 7 to 8) surroundings results from the simultaneous action of adsorption (Cu^{2+}, $Cu(OH)^+$) and also chemical deposition ($Cu(OH)_2^0$) [29,42,43].

The influence of pH on the extent of Cr(VI) uptake by oxMWCNT3h and oxMWCNT6h demonstrated an opposite effect to that described above. Namely, the capacity of oxMWCNTs for Cr(VI) at first increases, reaching a maximum at pH = 3, beyond which it begins to gradually decline. According to the speciation graph (Figure 8c), it is clear that in the highly acidic solutions (pH < 4) the most stable form of Cr(VI) is $HCrO_4^-$. These negatively charged oxyanions are readily electrostatically attracted and very likely chemically bonded (conclusion made on the basis of D–R isotherm model E_a values) to the oxMWCNTs surfaces (protonated and electrophilic centers), which are positive when pH < pHpzc, mainly as a consequence of the protonation of various weakly acidic and basic oxygen-containing functional groups (Equation (8)).

$$C_x - OH_2^+ + HCrO_4^- \leftrightarrow [C_x - OH_2CrO_4]^- + H^+ \tag{8}$$

The removal of Cr(VI) at low pH values can also take place through the occurrence of another ("indirect") mechanism, involving a three-step process: 1) reduction of $HCrO_4^-$ to Cr(III) by oxMWCNTs electron donors (various carboxyl and hydroxyl) groups. This reaction is maintained thanks to the high degree of Cr(VI) uptake at low pH; and the slow kinetics of this reaction are most likely responsible for the low Cr(VI) adsorption rate (Equation (9)), 2) release of Cr(III) species back into solution, 3) capturing Cr(III) aqua cations (mainly Cr^{3+}, speciation shown in Figure 8d) through the action of electrostatic attraction, chemisorption (deprotonated acidic sites) (Equation (10)) and ion exchange (protonated weakly acidic and basic sites) (Equation (11)). The $H^+/Cr(III)$ competition diminishes the effectiveness of this pathway, which is the reason why the total adsorption of Cr rises as pH goes up from pH 2 to pH 3.

$$3Cx - OH + HCrO_4^- + 4H^+ + 3e^- \leftrightarrow 3Cx - O^- + Cr^{3+} + 4H_2O \tag{9}$$

$$2C_x - CO^- + Cr^{3+} \leftrightarrow [(C_x - CO)_2Cr]^+ \tag{10}$$

$$2C_x - OH_2^+ + Cr^{3+} \leftrightarrow [(C_x - O)_2Cr]^{2+} + 2H^+ \tag{11}$$

It was not possible to confirm the presence of this mechanism via determination of the oxidation state of Cr (in solution or on the surfaces of oxMWCNT3h and oxMWCNT6h) after the adsorption process. However, the kinetic parameters, as well as the results of other authors who examined similar materials, clearly indicate its presence is likely [30].

At pH values greater than 3, deprotonation intensifies, the surface of the adsorbents becomes increasingly negative, and it becomes significantly more difficult for $HCrO_4^-$ to adsorb, due to electrostatic repulsion, according to Equation (8). This phenomenon is particularly pronounced in the case of higher charge number $Cr_2O_7^{2-}$ ions, the dominant form of Cr(VI) at pH > 6. Additionally, adsorption of oxyanions itself contributes to the negativity of oxMWCNTs surface, thus inhibiting the further removal of this metal. Finally, the existence of competition between excess OH^- and $HCrO_4^-/Cr_2O_7^{2-}$ for the same adsorption sites also has an adverse effect on the process studied [44].

3.5. Influence of Ionic Strength

The influence of the background electrolyte concentration on the degree of Cu(II), Ni(II) and Cr(VI) removal by oxMWCNT3h and oxMWCNT6h is illustrated in Figure 9. This is one of the potential controlling factors in any adsorption system.

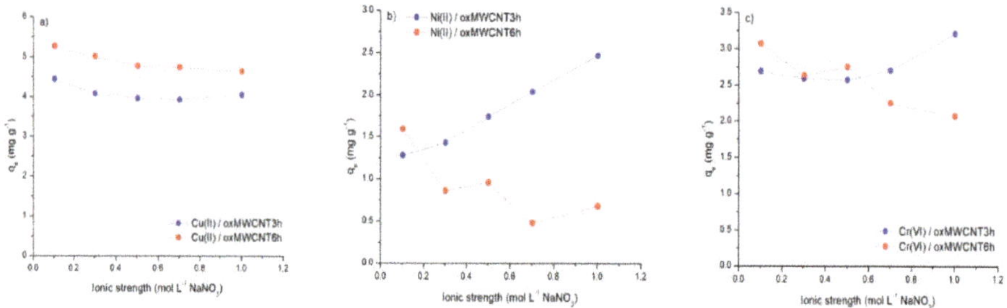

Figure 9. Adsorption of: (**a**) Cu(II), (**b**) Ni(II) and (**c**) Cr(VI) on oxMWCNT3h and oxMWCNT6h as a function of ionic strength (m = 5 mg, V = 30 mL (0.1–1 M NaNO$_3$), C_0 = 1 mg L^{-1}, pH = 5 ± 0.1 (Cu(II) and Ni(II)) and 2.5 ± 0.1 (Cr(VI)), t = 24 h (Cu(II) and Ni(II)) and 96 h (Cr(VI)), agitation speed = 180 rpm, T = 298 ± 2 K).

As can be seen, increasing the ionic strength from 0.1 to an extremely high level of 1 M NaNO$_3$, has a slight effect on the adsorption systems investigated, although it is considerably less than the effect of changing the solution pH. The observed variations in q_e (and RE_e) values are not completely consistent, even in the same metal/different adsorbent cases, and are located within the following limits: Cu(II)—3.93 to 4.44 mg g^{-1} (78.1% to 88.8%) and 4.64 to 5.27 mg g^{-1} (83.2% to 94.6%), Ni(II)—1.28 to 2.47 mg g^{-1} (25.4% to 48.5%) and 0.48 to 1.59 mg g^{-1} (8.5% to 28.1%), Cr(VI)—2.57 to 3.21 mg g^{-1} (48.0% to 60.0%) and 2.07 to 3.07 mg g^{-1} (34.7% to 51.2%). From the data reported above, it is clear that the greatest impact, be it positive or negative, depends on whether oxMWCNT3h or oxMWCNT6h is the adsorbent, in the case of adsorption of Ni(II).

Generally, when a solid charged surface is brought into contact with the liquid ionic medium, two layers of opposite polarity appear at the common boundary (interface) among the adjacent phases. Since the thickness of this electric double layer, as well as the interface potential, is determined by the background electrolyte concentration, the dominant types of interactions which take place on the related adsorbate/adsorbent exterior can be predicted.

According to the triple-layer (surface complexation) model, the uptake of Cu(II), Ni(II) and Cr(VI) associated with the oxMWCNTs surface, can occur in two different ways. The first of these involves the formation of inner-sphere complexes, that is partially or completely dehydrated metal ions directly and chemically specifically attached to adsorbent active sites. Such coordinatively bound species are located in the so-called o-plane. The second one, designated as outer-sphere complexation, is realized through the occurrence of electrostatic attraction between fully solvated metal ions, placed in the β plane on a favorably charged surface. Bearing in mind that the outer-sphere ion-pairs are set in the same region as background electrolyte counterions, it is expected for them to be more susceptible to ionic strength variations than strongly adsorbed o-plane complexes [45].

Considering all of the above, the large increase in Na^+/NO_3^- concentration required to cause only narrow changes in q_e values, when taken into account together with the conclusions on the high pH-dependence of Cu(II), Ni(II) and Cr(VI) adsorption, clearly indicate that the inner-sphere surface complexation (chemisorption) can be designated as the dominant process in the removal of these heavy metals from solution. Outer-sphere complexes can therefore be ruled out, which means the role of background ions competition must be excluded from the explanation of the slight reduction in metals uptake at high Na^+/NO_3^- concentrations. Instead, this reduction is explained as follows: 1) a high ionic strength environment lowers the repulsive electrostatic energy of oxMWCNTs, which results in solid particles agglomeration (destabilization) and therefore a decrease in the number of available active sites; 2) introduction of Na^+/NO_3^- decreases the activity coefficients of Cu(II), Ni(II) and Cr(VI) ions, thus retarding their transfer to the adsorbent's surface. The positive influence of ionic strength, evident only when it comes to the Ni(II)/oxMWCNT3h system, may be attributed to diminishing of some repulsive electrostatic interactions, possibly as a consequence of developing the screening effect. This finding cannot be considered very reliable, especially due to the fact that Ni(II) behaves completely differently in the presence of oxMWCNT6h, which according to the characterization results, is fairly similar to oxMWCNT3h. The observations derived from these experiments do agree with and further confirm those made in the previous sections [19,30,46,47].

4. Conclusions

Taking into account all the results obtained, the following conclusions can be drawn:

The Cu(II) and Ni(II) adsorption systems reach a state of equilibrium very quickly, in less than 20 min. This indicates that the removal of these metals is mainly determined by the interactions occurring at the adsorbate/adsorbent surface. The Cr(VI) uptake rate, divided into two stages, is much slower, with 96 h required to achieve equilibrium. The duration of the MWCNTs oxidation treatment does not influence the kinetics of these adsorption processes (oxMWCNT3h = oxMWCNT3h; Cu(II) = Ni(II) > Cr(VI)).

The Cr(VI) adsorption kinetics are well described by the non-linear Elovich equation, which is valid for mechanisms dominated by chemisorption. According to the linearization of Weber–Morris plots, intra-particle diffusion may be involved, but is not a rate determining factor in Cr(VI) removal by oxMWCNT3h and oxMWCNT6h. Instead, the overall rate and mechanism of adsorption in this systems are controlled by complex surface interactions, the most important of which, in this respect, relates to very slow Cr(VI)/oxMWCNTs redox reactions.

The Freundlich, Langmuir and D–R isotherms are all in good compliance with the corresponding experimental data. Values of n, R_L and E_a indicate that the Cu(II), Ni(II) and Cr(VI) adsorption on the applied MWCNTs is favorable and chemically specific. The capacity of oxMWCNT3h and oxMWCNT6h for these metals expressed as K_F, the parameter most consistent with the non-fitted observations, follows the order: Cu(II) > Cr(VI) > Ni(II) (Cu(II) and Ni(II): oxMWCNT6h > oxMWCNT3h, Cr(VI): oxMWCNT3h $\geq$ oxMWCNT6h).

All the studied systems are strongly influenced by the solution pH. Specifically, Cu(II) uptake increases as the initial pH increases from 2 to 6, Ni(II) removal rises throughout the entire pH range investigated, while Cr(VI) is best adsorbed at pH = 3. Ionic strength has a far less pronounced effect on the behavior of the adsorption processes investigated. These findings further confirm that the most important factor in the complicated mechanism of Cu(II), Ni(II) and Cr(VI) adsorption on oxMWCNT3h and oxMWCNT6h is inner-sphere surface complexation (chemisorption) involving the coordination of metal ions with oxygen-containing functional groups.

Supplementary Materials: The following are available online at http://www.mdpi.com/2073-4441/12/3/723/s1, Figure S1: SEM images of: (a) oxMWCNT3h and (b) oxMWCNT6h., Figure S2: Points of zero charge of: (**a**) oxMWCNT3h and (**b**) oxMWCNT6h. Table S1: EDS analysis of oxMWCNT3h and oxMWCNT6h.

Author Contributions: Conceptualization, S.M. and M.Š.; methodology, S.M., M.Š. and M.K.I.; software, M.W.; validation, S.M., M.W. and M.K.I.; formal analysis, J.N.; investigation; resources, J.T. and Z.K.; data curation, M.Š.; writing—original draft preparation, M.Š.; writing—review and editing, S.M.; visualization, M.Š.; supervision, S.M.; project administration, J.T.; funding acquisition, J.T. All authors have read and agreed to the published version of the manuscript.

Funding: This research was funded by the Ministry of Education, Science and Technological Development of the Republic of Serbia (Project No. 451-03-68/2020-14/ 200125).

Acknowledgments: The authors gratefully acknowledge the support of the Ministry of Education, Science and Technological Development of the Republic of Serbia (Project No. 451-03-68/2020-14/ 200125).

Conflicts of Interest: The authors declare no conflict of interest.

References

1. Vareda, J.P.; Valente, A.J.M.; Durães, L. Assessment of heavy metal pollution from anthropogenic activities and remediation strategies: A review. *J. Environ. Manag.* **2019**, *246*, 101–118. [CrossRef] [PubMed]

2. Vardhan, K.H.; Kumar, P.S.; Panda, R.C. A review on heavy metal pollution, toxicity and remedial measures: Current trends and future perspectives. *J. Mol. Liq.* **2019**, *290*, 111197. [CrossRef]

3. Panagos, P.; Van Liedekerke, M.; Yigini, Y.; Montanarella, L. Contaminated Sites in Europe: Review of the Current Situation Based on Data Collected through a European Network. *Ecol. Indic.* **2013**, *24*, 439–450. [CrossRef]

4. Kumari, P.; Alam, M.; Siddiqi, W.A. Usage of nanoparticles as adsorbents for waste water treatment: An emerging trend. *Sustain. Mater. Technol.* **2019**, *22*, e00128. [CrossRef]

5. Yadav, V.B.; Gadi, R.; Kalra, S. Clay based nanocomposites for removal of heavy metals from water: A review. *J. Environ. Manag.* **2019**, *232*, 803–817. [CrossRef]

6. WHO. *Guidelines for Drinking-Water Quality*, 4th ed.; WHO: Geneva, Switzerland, 2012.

7. EU. European Commission Implementing Decision (EU) 2017/1442 of 31 July 2017 establishing best available techniques (BAT) conclusions, under Directive 2010/75/EU of the European Parliament and of the Council, for large combustion plants. *Off. J. Eur. Union* **2017**. Available online: http://www.legislation.gov.uk/eudn/2017/1442 (accessed on 5 March 2020).

8. EU. European Commission Implementing Decision (EU) 2016/1032 of 13 June 2016 establishing best available techniques (BAT) conclusions, under Directive 2010/75/EU of the European Parliament and of the Council, for the non-ferrous metals industries. *Off. J. Eur. Union* **2016**, 32–106. Available online: http://www.legislation.gov.uk/eudn/2016/1032 (accessed on 5 March 2020).

9. Nasir, A.M.; Goh, P.S.; Abdullah, M.S.; Ng, B.C.; Ismail, A.F. Adsorptive nanocomposite membranes for heavy metal remediation: Recent progresses and challenges. *Chemosphere* **2019**, *232*, 96–112. [CrossRef]

10. Xu, J.; Cao, Z.; Zhang, Y.; Yuan, Z.; Lou, Z.; Xu, X.; Wang, X. A review of functionalized carbon nanotubes and graphene for heavy metal adsorption from water: Preparation, application, and mechanism. *Chemosphere* **2018**, *195*, 351–364. [CrossRef]

11. Joseph, L.; Jun, B.-M.; Flora, J.R.V.; Park, C.M.; Yoon, Y. Removal of heavy metals from water sources in the developing world using low-cost materials: A review. *Chemosphere* **2019**, *229*, 142–159. [CrossRef]

12. Wang, L.; Wang, Y.; Ma, F.; Tankpa, V.; Bai, S.; Guo, X.; Wang, X. Mechanisms and reutilization of modified biochar used for removal of heavy metals from wastewater: A review. *Sci. Total Environ.* **2019**, *668*, 1298–1309. [CrossRef]

13. Fiyadh, S.S.; AlSaadi, M.A.; Jaafar, W.Z.; AlOmar, M.K.; Fayaed, S.S.; Mohd, N.S.; Hin, L.S.; El-Shafie, A. Review on heavy metal adsorption processes by carbon nanotubes. *J. Clean. Prod.* **2019**, *230*, 783–793. [CrossRef]

14. Gupta, N.K.; Choudhary, B.C.; Gupta, A.; Achary, S.N.; Sengupta, A. Graphene-based adsorbents for the separation of f-metals from waste solutions: A review. *J. Mol. Liq.* **2019**, *289*, 111121. [CrossRef]

15. Joseph, L.; Jun, B.-M.; Jang, M.; Park, C.M.; Muñoz-Senmache, J.C.; Hernández-Maldonado, A.J.; Heyden, A.; Yu, M.; Yoon, Y. Removal of contaminants of emerging concern by metal-organic framework nanoadsorbents: A review. *Chem. Eng. J.* **2019**, *369*, 928–946. [CrossRef]

16. Almeida, J.C.; Cardoso, C.E.D.; Tavares, D.S.; Freitas, R.; Trindade, T.; Vale, C.; Pereira, E. Chromium removal from contaminated waters using nanomaterials—A review. *TrAC Trends Anal. Chem.* **2019**, *118*, 277–291. [CrossRef]

17. Farghali, A.A.; Abdel Tawab, H.A.; Abdel Moaty, S.A.; Khaled, R. Functionalization of acidified multi-walled carbon nanotubes for removal of heavy metals in aqueous solutions. *J. Nanostruct. Chem.* **2017**, *7*, 101–111. [CrossRef]

18. Mobasherpour, I.; Salahi, E.; Ebrahimi, M. Removal of divalent nickel cations from aqueous solution by multi-walled carbon nano tubes: Equilibrium and kinetic processes. *Res. Chem. Intermed.* **2012**, *38*, 2205–2222. [CrossRef]

19. Ge, Y.; Li, Z.; Xiao, D.; Xiong, P.; Ye, N. Sulfonated multi-walled carbon nanotubes for the removal of copper (II) from aqueous solutions. *J. Ind. Eng. Chem.* **2014**, *20*, 1765–1771. [CrossRef]

20. Abdel-Ghani, N.T.; El-Chaghaby, G.A.; Helal, F.S. Individual and competitive adsorption of phenol and nickel onto multiwalled carbon nanotubes. *J. Adv. Res.* **2015**, *6*, 405–415. [CrossRef]

21. Zhao, X.H.; Jiao, F.P.; Yu, J.G.; Xi, Y.; Jiang, X.Y.; Chen, X.Q. Removal of Cu(II) from aqueous solutions by tartaric acid modified multi-walled carbon nanotubes. *Colloids Surfaces A Physicochem. Eng. Asp.* **2015**, *476*, 35–41. [CrossRef]

22. Kandah, M.I.; Meunier, J.L. Removal of nickel ions from water by multi-walled carbon nanotubes. *J. Hazard. Mater.* **2007**, *146*, 283–288. [CrossRef]

23. Kosa, S.A.; Al-Zhrani, G.; Abdel Salam, M. Removal of heavy metals from aqueous solutions by multi-walled carbon nanotubes modified with 8-hydroxyquinoline. *Chem. Eng. J.* **2012**, *181–182*, 159–168. [CrossRef]

24. Kanyó, T.; Kónya, Z.; Kukovecz, Á.; Berger, F.; Dékány, I.; Kiricsi, I. Quantitative characterization of hydrophilic-hydrophobic properties of MWNTs surfaces. *Langmuir* **2004**, *20*, 1656–1661. [CrossRef]

25. Abdel Salam, M.; Al-Zhrani, G.; Kosa, S.A. Removal of heavy metal ions from aqueous solution by multi-walled carbon nanotubes modified with 8-hydroxyquinoline: Kinetic study. *J. Ind. Eng. Chem.* **2014**, *20*, 572–580. [CrossRef]

26. Ayawei, N.; Ebelegi, A.N.; Wankasi, D. Modelling and Interpretation of Adsorption Isotherms. *J. Chem.* **2017**, *2017*, 3039817. [CrossRef]

27. Kragulj, M.; Tričković, J.; Dalmacija, B.; Kukovecz, Á.; Kónya, Z.; Molnar, J.; Rončević, S. Molecular interactions between organic compounds and functionally modified multiwalled carbon nanotubes. *Chem. Eng. J.* **2013**, *225*, 144–152. [CrossRef]

28. Wang, S.; Sun, H.; Ang, H.M.; Tade, M.O. Adsorptive remediation of environmental pollutants using novel graphene-based nanomaterials. *Chem. Eng. J.* **2013**, *226*, 336–347. [CrossRef]

29. Yang, S.; Li, J.; Shao, D.; Hu, J.; Wang, X. Adsorption of Ni(II) on oxidized multi-walled carbon nanotubes: Effect of contact time, pH, foreign ions and PAA. *J. Hazard. Mater.* **2009**, *166*, 109–116. [CrossRef] [PubMed]

30. Hu, J.; Chen, C.; Zhu, X.; Wang, X. Removal of chromium from aqueous solution by using oxidized multiwalled carbon nanotubes. *J. Hazard. Mater.* **2009**, *162*, 1542–1550. [CrossRef]

31. Gholipour, M.; Hashemipour, H. Evaluation of multi-walled carbon nanotubes performance in adsorption and desorption of hexavalent chromium. *Chem. Ind. Chem. Eng. Q.* **2012**, *18*, 509–523. [CrossRef]

32. Can, M. Studies of the kinetics for rhodium adsorption onto gallic acid derived polymer: The application of nonlinear regression analysis. *Acta Phys. Pol. A* **2015**, *127*, 1308–1310. [CrossRef]

33. Lee, C.; Kim, S. Cr(VI) Adsorption to Magnetic Iron Oxide Nanoparticle-Multi-Walled Carbon Nanotube Adsorbents. *Water Environ. Res.* **2016**, *88*, 2111–2120. [CrossRef]

34. Tran, H.N.; You, S.J.; Hosseini-Bandegharaei, A.; Chao, H.P. Mistakes and inconsistencies regarding adsorption of contaminants from aqueous solutions: A critical review. *Water Res.* **2017**, *120*, 88–116. [CrossRef] [PubMed]

35. Jung, C.; Heo, J.; Han, J.; Her, N.; Lee, S.J.; Oh, J.; Ryu, J.; Yoon, Y. Hexavalent chromium removal by various adsorbents: Powdered activated carbon, chitosan, and single/multi-walled carbon nanotubes. *Sep. Purif. Technol.* **2013**, *106*, 63–71. [CrossRef]

36. Tofighy, M.A.; Mohammadi, T. Adsorption of divalent heavy metal ions from water using carbon nanotube sheets. *J. Hazard. Mater.* **2011**, *185*, 140–147. [CrossRef] [PubMed]

37. Zhang, X.; Huang, Q.; Liu, M.; Tian, J.; Zeng, G.; Li, Z.; Wang, K.; Zhang, Q.; Wan, Q.; Deng, F.; et al. Preparation of amine functionalized carbon nanotubes via a bioinspired strategy and their application in Cu^{2+} removal. *Appl. Surf. Sci.* **2015**, *343*, 19–27. [CrossRef]

38. Labied, R.; Benturki, O.; Eddine Hamitouche, A.Y.; Donnot, A. Adsorption of hexavalent chromium by activated carbon obtained from a waste lignocellulosic material (Ziziphus jujuba cores): Kinetic, equilibrium, and thermodynamic study. *Adsorpt. Sci. Technol.* **2018**, *36*, 1066–1099. [CrossRef]

39. Yu, F.; Wu, Y.; Ma, J.; Zhang, C. Adsorption of lead on multi-walled carbon nanotubes with different outer diameters and oxygen contents: Kinetics, isotherms and thermodynamics. *J. Environ. Sci. (China)* **2013**, *25*, 195–203. [CrossRef]

40. Burakov, A.E.; Galunin, E.V.; Burakova, I.V.; Kucherova, A.E.; Agarwal, S.; Tkachev, A.G.; Gupta, V.K. Adsorption of heavy metals on conventional and nanostructured materials for wastewater treatment purposes: A review. *Ecotoxicol. Environ. Saf.* **2018**, *148*, 702–712. [CrossRef]

41. Sarma, G.K.; Sen Gupta, S.; Bhattacharyya, K.G. Nanomaterials as versatile adsorbents for heavy metal ions in water: A review. *Environ. Sci. Pollut. Res.* **2019**, *26*, 6245–6278. [CrossRef]

42. Sun, X.; Liu, X.; Yang, B.; Xu, L.; Yu, S. Functionalized chrysotile nanotubes with mercapto groups and their Pb(II) and Cd(II) adsorption properties in aqueous solution. *J. Mol. Liq.* **2015**, *208*, 347–355. [CrossRef]

43. Xiao, D.L.; Li, H.; He, H.; Lin, R.; Zuo, P.L. Adsorption performance of carboxylated multi-wall carbon nanotube-Fe3O4 magnetic hybrids for Cu(II) in water. *Xinxing Tan Cailiao/New Carbon Mater.* **2014**, *29*, 15–25. [CrossRef]

44. Di Natale, F.; Lancia, A.; Molino, A.; Musmarra, D. Removal of chromium ions form aqueous solutions by adsorption on activated carbon and char. *J. Hazard. Mater.* **2007**, *145*, 381–390. [CrossRef] [PubMed]

45. Sheng, G.; Li, J.; Shao, D.; Hu, J.; Chen, C.; Chen, Y.; Wang, X. Adsorption of copper(II) on multiwalled carbon nanotubes in the absence and presence of humic or fulvic acids. *J. Hazard. Mater.* **2010**, *178*, 333–340. [CrossRef] [PubMed]

46. Chen, C.; Hu, J.; Shao, D.; Li, J.; Wang, X. Adsorption behavior of multiwall carbon nanotube/iron oxide magnetic composites for Ni(II) and Sr(II). *J. Hazard. Mater.* **2009**, *164*, 923–928. [CrossRef] [PubMed]

47. López-Ramón, V.; Moreno-Castilla, C.; Rivera-Utrilla, J.; Radovic, L.R. Ionic strength effects in aqueous phase adsorption of metal ions on activated carbons. *Carbon N. Y.* **2003**, *41*, 2020–2022. [CrossRef]

Article

Arsenic Removal from Water by Green Synthesized Magnetic Nanoparticles

Jasmina Nikić, Aleksandra Tubić *, Malcolm Watson, Snežana Maletić, Marko Šolić, Tatjana Majkić and Jasmina Agbaba

Department of Chemistry, Biochemistry and Environmental Protection, Faculty of Sciences, University of Novi Sad, Trg Dositeja Obradovića 3, 21000 Novi Sad, Serbia; jasmina.nikic@dh.uns.ac.rs (J.N.); malcolm.watson@dh.uns.ac.rs (M.W.); snezana.maletic@dh.uns.ac.rs (S.M.); marko.solic@dh.uns.ac.rs (M.Š.); tatjana.majkic@dh.uns.ac.rs (T.M.); jasmina.agbaba@dh.uns.ac.rs (J.A.)
* Correspondence: aleksandra.tubic@dh.uns.ac.rs

Received: 30 September 2019; Accepted: 25 November 2019; Published: 28 November 2019

Abstract: Magnetite nanoparticles were synthesized by a simple and ecofriendly method using onion peel (MNp-OP) and corn silk extract (MNp-CS), in order to develop new low-cost adsorbents for arsenic removal from groundwater. As a point of comparison, magnetite nanoparticles were also synthesized with a conventional chemical process (MNp-CO). The antioxidant potential of onion peel and corn silk extracts was determined using ferric reducing antioxidant power (FRAP) and free radical (DPPH) scavenging assays, including the total phenolics, flavonoids and tannins contents. The synthesized magnetite nanoparticles were characterised using different techniques (Scanning electron microscope/Energy dispersive spectroscopy (SEM/EDS), X-ray diffraction (XRD), Fourier-transform infrared spectroscopy (FTIR) and Brunauer-Emmett-Teller (BET) surface area analyzer). The adsorption capacity of MNp-OP and MNp-CS and the arsenic removal mechanism of these novel adsorbents was investigated through kinetic and equilibrium experiments and their corresponding mathematical models. Characterisation of MNp-OP and MNp-CS shows high BET specific surface areas of 243 m^2/g and 261 m^2/g, respectively. XRD and FTIR analysis confirmed the formation and presence of magnetite nanoparticles. The arsenic adsorption mechanism on MNp-OP, MNp-CS and MNp-CO involves chemisorption, intraparticle and external diffusion. Maximal adsorption capacities of MNp-OP, MNp-CS and MNp-CO were 1.86, 2.79, and 1.30 mg/g respectively. The green synthesis applied using onion peel and corn silk extracts was cost effective and environmentally friendly, and results in adsorbents with a high capacity for arsenic and magnetic properties, making them a very promising alternative approach in the treatment of arsenic contaminated groundwater.

Keywords: magnetite nanoparticles; onion peel; corn silk; arsenic; adsorption; groundwater

---

1. Introduction

The presence of arsenic in groundwater, either from anthropogenic or natural sources, is a serious problem in many different parts of the world. Millions of people, mostly in rural and developing countries, are exposed to high levels of arsenic via the intake of arsenic rich groundwater [1]. The main concerns related to arsenic in groundwater are its high toxicity and carcinogenicity, since long-term exposure to arsenic has been associated with cancer (skin, lungs, urinary tract, kidneys and liver) and other various noncancerous diseases [2]. In order to minimize and reduce the adverse health effects, the World Health Organization (WHO) recommends a maximum allowable concentration (MAC) for arsenic in drinking water of 10 µg/L [3].

Various technologies have been employed for arsenic removal from groundwater including coagulation, membrane separation, ion exchange and adsorption [4]. Among these techniques, adsorption offers many advantages including simple and stable operation, easy handling of waste, absence of added reagents, compact facilities, and generally lower operation costs [5–8].

Many synthetic and natural adsorbents have been developed and applied for arsenic removal, however in recent years, iron nanoparticles (INPs) have been the most interesting novel materials, due to their unique physicochemical properties (small particle size, high surface area, high magnetism, low toxicity) and their strong affinity for arsenic species [9–12]. INPs can be classified into three major groups namely:

(1)   iron oxide nanoparticles (IONs) (i.e., magnetite; Fe_3O_4, hematite: α-Fe_2O_3, maghemite; γ-Fe_2O_3),
(2)   iron oxide hydroxide (FeOOH) nanoparticles and
(3)   zero-valent iron (ZVI) nanoparticles.

Among the iron oxide nanoparticles, magnetite (Fe_3O_4) and maghemite (γ-Fe_2O_3) nanoparticles, have attracted a lot of attention for applications with arsenic contaminated waters [13–16]. Their ease of fabrication and high adsorption capacities for arsenic and super-paramagnetic properties allow for their easy separation from water, make them popular candidates for remediation of arsenic contaminated groundwaters.

Different methods have been reported for the synthesis of magnetic nanoparticles such as sol–gel process [16], electrical wire explosion (EWE) [13], hydrothermal synthesis [17], coprecipitation [18], etc. These methods require toxic solvents, high energy consumption and/or generate hazardous by-products. In contrast to these chemical and physical methods, the green synthesis of nanoparticles has been proposed as a cost-effective environmentally friendly and promising approach [19]. Biogenic or green synthesis of magnetic nanoparticles includes the use of plants, algae, and different microorganisms (yeast, fungi, diatoms, and bacteria), with plant-mediated synthesis of magnetite nanoparticles being the most investigated. Different parts of plants can be applied for the synthesis of magnetic nanoparticles; however, most researchers apply various leaf extracts [20–25]. In general, the extracts obtained from plants contain a variety of phytochemicals (polyphenolics, flavonoid, tannins etc.) and various types of proteins, enzymes, polysaccharides and alcoholic compounds which can act as natural sources for reducing and capping agents in a one-pot synthesis reaction. Such an approach eliminates multistep synthesis practice problems and the costs of chemical reagents. Moreover, the use of plant residues from non-food sources (which would otherwise be waste materials) as a feedstock for green synthesis reactions reduces production costs and further contributes to sustainability.

To date, there is no literature relating to the green synthesis of magnetite nanoparticles using extracts obtained from onion peel and corn silk, or their application on arsenic remediation.

Onion (*Allium cepa* L.) is one of the most frequently consumed vegetables, and its production increases every day due to increasing consumer demand. Simultaneously, huge amounts of waste are produced from different parts of the onion, which affect the environment in various ways. Hence, proper use as well as disposal of this waste is important from an environmental aspect [26]. The concentrations of total phenols, flavonoids, flavonols and other antioxidants are higher in the onion peel than in the edible portion of the onion [27–30]. This waste therefore has great promise as a natural reductant in the one-pot synthesis of magnetic nanoparticles.

Corn silk, the female flower stigma of maize (*Zea mays* L.), is a yellow silky strand found on the top of corn fruit. Corn silk is a by-product obtained from corn production such as food, processed food and animal feed, and is discarded as waste together with other parts of corn such as the cob and husk [31]. Like onion peel, corn silk contains various components, including protein, vitamins, carbohydrates, Ca, K, Mg, sitosterol, stigmasterol, alkaloids, saponins, tannins and flavonoids (maysin, methoxymaysin, apimaysin, and luteolin derivatives). Moreover, it is a good source of polyphenol compounds, which are strong antioxidants [32,33]. Corn silk is used in traditional medicine to treat many diseases [34,35], but despite its medicinal properties, it is still not used in the pharmaceutical industry and is treated as a waste material.

Thus, in this study magnetite nanoparticles were synthesized for the first time in a one-step process using extract obtained from onion shell (Mag-OS) and corn silk (Mag-CS). These novel adsorbents were characterized by different techniques (SEM/EDS, XRD, FTIR, BET) and further applied for arsenic removal from groundwater. In order to compare efficiency of the green synthesis, as well arsenic adsorption capacities, magnetic nanoparticles were also synthesized by conventional chemical coprecipitation method (MNp-CO), the most common fabrication pathway for this material.

2. Materials and Methods

2.1. Green Synthesis of Magnetic Nanoparticles

2.1.1. Preparation of Onion Peel and Corn Silk Extracts

Onion peels (OP) and corn silk (CS) were collected from locally grown produce (Dutch yellow onions and NS 6010 Hybrid corn, both from NS Seme, Novi Sad). The collected materials were washed several times with distilled water to remove impurities, dried at ambient temperature, then cut into small pieces. Ten g of CS/OP were weighed and transferred to a 500 mL beaker, to which 300 mL of distilled water was added. The beaker was then placed in a shaker bath at 30 °C for 1 h. After shaking, the solution was filtered with a Büchner Vacuum Filtration Funnel and the filtrate was used for magnetite nanoparticles production.

2.1.2. Preparation of Magnetite Nanoparticles

The magnetic nanoparticles were synthesized by green precipitation method without addition of a chemical reducing agent. In green synthesis, the plant extracts act as chemical reducing agents while sodium acetate acts as an electrostatic stabilizing agent for the produced Fe_3O_4 nanoparticles [36]. In the reaction procedure, 2.16 g of $FeCl_3 \cdot 6H_2O$ and 6.56 g of sodium acetate was dissolved in 40 mL of corn silk/onion peel extract. The suspension was placed under vigorous stirring for 1 h at 80 °C, and the resulting precipitate centrifuged and washed several times with distilled water. The obtained magnetic nanoparticles were dried in an oven at 105 °C. The synthesized magnetic nanoparticles were denoted as MNp-CS and MNp-OP.

2.2. Synthesis of Magnetic Nanoparticles by Coprecipitation Methods

In order to compare the efficiency of the green synthesis method and the adsorption capacity of the green nanoparticles obtained (MNp-CS and MNp-OP), magnetic nanoparticles were also synthesized via the chemical coprecipitation method using Fe^{3+} and Fe^{2+} salts at a molar ratio of 2:1. 2.50 g of $FeCl_3 \cdot 6H_2O$ and 1.21 g $FeSO_4 \cdot 7H_2O$ were dissolved in 40 mL of deionized water in a 200 mL glass beaker. Two M NaOH solution was added dropwise until the mixture reached pH 9 and a black precipitate developed. The synthesized MNp-CO were then separated from the suspension by external magnet and washed repeatedly with deionised water until a supernatant of neutral pH was obtained.

2.3. Characteristion of Plant Extracts

Total phenolics in the plant extract were determined by Folin–Ciocalteu reagent (FC), as previously described [37]. The total phenol content (TPC) was expressed as mg of gallic acid equivalents (GAE) per g dry weight (dw) of extracts. The tannin content was determined using insoluble polyvinyl polypirrolidone, which binds the tannins. The total tannin content (TTC) was expressed as mg of catechin equivalents (CE) per g dry weight (dw) of extracts. The total content of flavonoids (TFC) was determined by using the aluminium chloride colorimetric method as previously described. The total content of flavonoids was expressed as mg of quercetin equivalents (QE) per g dry weight (dw) of extracts. The antioxidant potential of plant extracts was determined using tests related to free radical (DPPH$^\bullet$) scavenging.

Reducing power was measured by Ferric Reducing Antioxidant Power (FRAP) assay according to the previously described procedure [38]. Reducing power were expressed as mmol of trolox equivalents (TE) per g dry weight (dw) of extracts.

2.4. Characterisation of Magnetite Nanoparticles

The specific surface area of the magnetic nanoparticles was measured by nitrogen adsorption using the Brunauer–Emmett–Teller (BET) method with a Autosorb TMiQ surface area analyzer (Quantachrome, Boynton Beach, FL, USA). Mesopore and micropore volumes were determined using the Barett–Joyner–Halender (BJH) method using desorption isotherm, and *t*-test method, respectively. The surface morphology, microstructure and elemental composition of the magnetic nanoparticles were examined using scanning electron microscopy combined with energy dispersive X-ray spectroscopy (SEM-EDX) (Hitachi TM3030, Japan). X-ray diffraction (XRD) patterns were obtained on Philips PW automated X-ray powder diffractometer (USA). Fourier transform infrared (FTIR) spectra of magnetic nanoparticles were recorded by infrared spectrophotometer (FTIR Nexus 670, Thermonicolet, USA). The point of zero charge (pH_{pzc}) of the magnetic nanoparticles was determined according to the method described by Zhang et al. [39]

2.5. Adsorption Experiments

Adsorption experiments were conducted with groundwater samples taken from Višnjićevo (a settlement located in the Autonomous Province of Vojvodina, Serbia). The characteristics of the groundwater are given in Table 1. Measurement uncertainties are expressed as the standard deviations of 10 separate measurements. For the kinetics experiments, 20 mL water samples were added to 40 mL glass bottles containing 10 mg of synthesized magnetic nanoparticles (MNp-CS, MNp-OP and MNp-CO). The suspensions were shaken on an orbital shaker (180 rpm, 22 °C) for certain periods of time (1, 2, 4, 6, 9, 12, 18, 24 h) and afterwards the supernatants were separated from the adsorbents using an external magnet. Residual arsenic concentrations in the samples were analyzed by inductively coupled plasma mass spectrometry (ICP-MS) as explained below. Equilibrium experiments were conducted with different sorbent doses (0.01–0.1 g) in the same manner as the kinetic experiments.

Table 1. Characteristics of the investigated groundwater.

Parameter	Value ± sd [a]
pH	8.05 ± 0.15
Conductivity (μs/cm)	669 ± 14.3
Alkalinity (mmol/L)	6.93 ± 0.37
Hardness (mg $CaCO_3$/L)	133 ± 48.3
Total arsenic (μg/L)	126 ± 8.97
Arsenic (III)	98 ± 12
Arsenic (V)	18 ± 2.5
Fe (μg/L)	20.2 ± 25.8
TOC (mgC/L)	2.40 ± 0.71
DOC (mgC/L)	2.12 ± 0.51
Phosphate (mgPO_4/L)	1.33 ± 0.05
Chloride (mgCl/L)	17.7 ± 1.36
Ammonium (mgN/L)	0.15 ± 0.10
Sulfate (mgSO_4/L)	20.3 ± 12.2

[a] sd-standard deviation, based on 10 measurements.

The amount of arsenic adsorbed by a unit mass of adsorbent was calculated using the following equation:

$$q_e = \frac{(C_0 - C_e)\,V}{m} \tag{1}$$

where C_0 and C_e are the initial and equilibrium concentrations in solution (mg/L). V is the volume of solutions (L) and m the mass of sorbent applied during the experiment (g).

Different kinetic models (pseudo-first order, pseudo-second order model, Elovich, Weber–Morris and Boyd model) [40] were used to fit the kinetic experimental data while the Langmuir, Freundlich, Dubinin–Radushkevich and Tempkin isotherm [41] were applied for modelling equilibrium experimental data. The parameters of these models were determined by nonlinear fitting of the data using Origin 8.0 software (OriginLab Corporation, Northampton, MA, USA).

2.6. Analytical Methods

pH measurements were carried out using an InoLab pH/ION 735 instrument (WTW, Austria). Total arsenic concentrations were analyzed by ICPMS (7700 Series, Agilent Technologies, Tokyo, Japan) and arsenic speciation analyses were carried out on the same system coupled with high performance liquid chromatography (1260 Infinity, Agilent Technologies, Germany) [42,43]. Water alkalinity (p- and m-alkalinity) was measured by titration with a standard solution of HCl according to Standard Methods [44]. The concentration of orthophosphate was determined according to the official Serbian translation of a method published by the International Organization for Standardization (SRPS EN ISO 6878: 2008) [45]. Concentration of chloride and ammonium was determined according to methods SRPS ISO 9297/1:2007 and SRPS ISO 5664:1992, respectively [46,47]. Dissolved organic carbon (DOC) in groundwater was analysed after filtration through a 0.45 μm membrane filter on an LiquiTOCII (Elementar, Germany), using Pt catalysed combustion at 850 °C to oxidize the carbon, in accordance with standard method SPRS ISO 8245:2007 [48].

3. Results and Discussion

3.1. Characterization of Magnetic Nanoparticles

The addition of $FeCl_3$ to the OP and CS extracts initiated the formation of magnetite nanoparticles MNp-OP and MNp-CS. The surface morphology, as well as the qualitative and semiquantitative composition of the surface of these nanoparticles, was investigated by SEM/EDS analysis, as shown in Figure 1. The magnetite nanoparticles (MNp-CO) synthesized for the sake of comparison by conventional coprecipitation method is also shown.

It can be seen that the surfaces of all three materials are rough and contain a large number of particles that are irregular in shape and size. A large number of aggregated particles on the surface of these materials could be attributed to nanostructures and magnetic properties, which contribute a tendency for agglomeration and aggregation.

EDS analysis of MNp-OP showed iron and oxygen weights of 57.1% and 40.3%, respectively. Similarly, MNp-CS contained 55.6% Fe and 40.9% O. Trace amounts of Cl and C were found on the surface of both materials (<2%). As with MNp-OP and MNp-CS, EDS analysis of MNp-CO showed a molar ratio of Fe:O of 41.7% to 43.2%. In general, the presence of Fe and O peaks in the surface of these materials confirms the formation of iron oxide. The presence of carbon can be attributed either to polyphenols or other phytochemical compounds or CO_2 from air. The traces of Cl come from the $FeCl_3$ precursor.

The phase identification and crystalline structures of the nanoparticles was characterized by X-ray powder diffraction. The X-ray diffraction patterns of MNp-OP, MNp-CS and MNp-CO are presented in Figure 2.

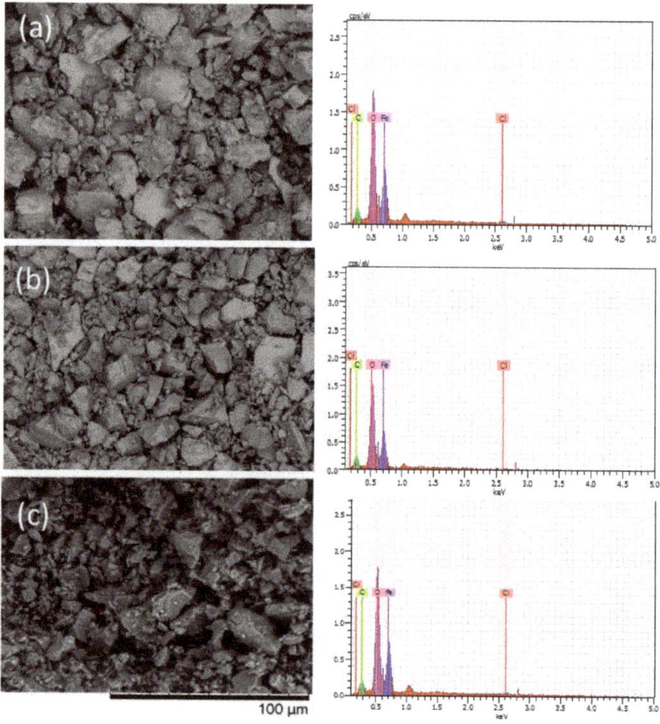

Figure 1. Scanning electron microscope (SEM) and energy-dispersive spectroscopy (EDS) images of (**a**) MNp-OP (**b**) MNp-CS (**c**) MNp-CO.

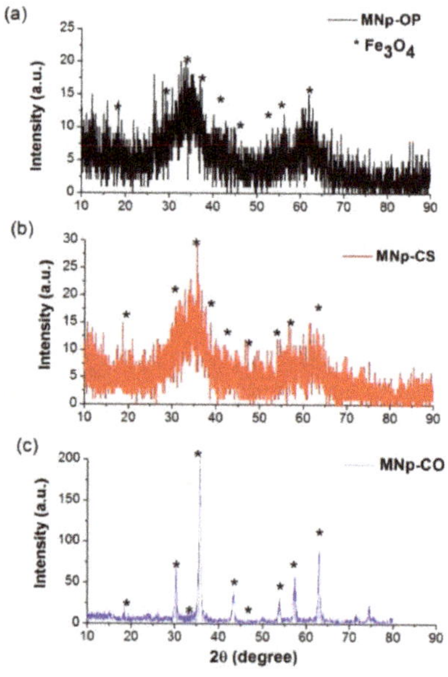

Figure 2. X-ray diffraction (XRD) patterns of (**a**) MNp-OP (**b**) MNp-CS and (**c**) MNp-CS.

XRD patterns show that all diffraction peaks at 2θ = 18.7, 30.4, 35.5, 37.7, 43.8, 47.3, 54.2, 56.8 and 62.5, show good matches with crystal planes of magnetite at (111), (220), (311), (222), (400), (331), (422), (511) and (440). Additionally, it may be noted that the peaks in XRD patterns of these materials were broad with low intensity, indicating nanosized particles [25,36,49]. The average particle sizes of the synthesized MNp-OP, MNp-CS and MNp-CO, which were calculated using the Debye–Scherrer equation [50] were found to be 26 nm, 28 nm and 12 nm, respectively. Thus, the XRD patterns indicate that the magnetic nanoparticles were successfully synthesized using the green method.

The magnetic nanoparticles were also analyzed by FTIR spectroscopy. The corresponding FTIR spectra of MNp-OP, MNp-CS and MNp-CO are presented in Figure 3.

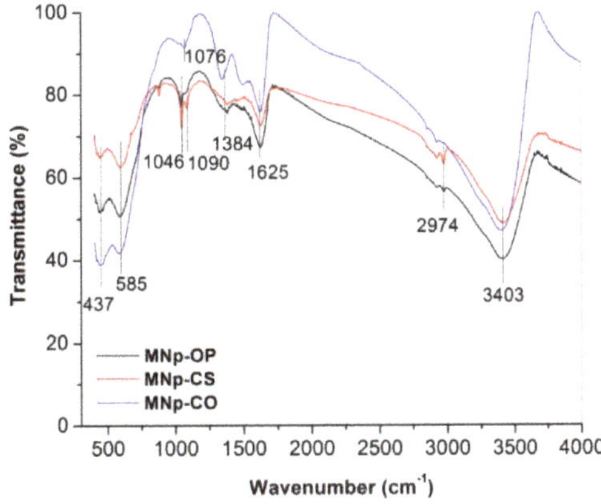

Figure 3. Fourier-transform infrared spectroscopy (FTIR) spectra of MNp-OP, MNp-CS and MNp-CO.

All spectra exhibited broad and strong bands at 3420 cm^{-1} and 1630 cm^{-1}, attributable to O–H stretching and O–H bending vibration from physisorbed water molecules [51]. The band at 2974 cm^{-1} is attributed to the C–H stretching vibration of hydrocarbon chains. The peaks at 1046, 1076 and 1090 cm^{-1} correspond to the bending vibration of the hydroxyl groups (Fe–OH), which are responsible for the formation of inner sphere surface complexes [52]. Absorption peaks at 585 cm^{-1} and 437 cm^{-1} are characteristic peaks for magnetite confirming that green synthesis as a chemical synthesis can produce Fe$_3$O$_4$ nanoparticles [25,36,49].

The textural characteristics of the synthesized magnetic nanoparticles, including specific surface area, mesopore and micropore volumes and average pore size of the synthesized sorbents, obtained by BET, BJH and model *t*-tests, are shown in Table 2.

Table 2. Characteristics of MNp-OP, MNp-CS and MNp-CO.

Adsorbents	Specific Surface Area (m^2/g)	Mesopore Volume (cm^3/g)	Micropore Volume (cm^3/g)	Average Pore Size (nm)
MNp-OP	243	0.067	0.038	1.42
MNp-CS	261	0.066	0.042	1.29
MNp-CO	72.1	0.155	0	10.5

BET analysis show that the specific surface area of magnetic nanoparticles synthesized via the green method, MNp-OP and MNp-CS, have much larger specific surfaces than MNp-CO and other structurally similar adsorbents, such as magnetite nanoparticles synthesized with plantain peel extract (11.31 m^2/g) [36], commercially available magnetite and maghemite nanoparticles (40 and 39 m^2/g, respectively), magnetite nanoparticles synthesized via electrical wire explosion (EWE) (12 m^2/g) [13], etc. The mesopore volume of MNp-CO obtained by BJH method was higher than for green synthesized magnetic nanoparticles. However, in comparison with MNp-CO, the *t*-test method shows that MNp-OP and MNp-CS also contain micropores (0.038 and 0.042 cm^3/g, respectively) and pore size less than 2 nm, which means that according to the International Union of Pure and Applied Chemistry (IUPAC) classification, these materials may be considered as microporous adsorbents (pore diameter less than 2 nm) [53]. The magnetite nanoparticles synthesized via conventional precipitation method, MNp-CO, do not have micropores, and its pore size was 10.5 nm, implying a mesoporous structure of this material (pore diameter in the range 2 to 50 nm).

The points of zero charge of MNp-OP, MNp-CS and MNp-CO are pH_{pzc} = 5.63, pH_{pzc} = 6.75 and pH_{pzc} = 6.18, respectively (Figure 4). It is well known that at pH values less than pH_{pzc}, the surface of the sorbent is positively charged, which favours the sorption of negatively charged arsenic anions. In contrast, at pH values higher than pH_{pzc}, the surface of the sorbent is negatively charged, leading to electrostatic repulsions between the surface of the sorbent and the arsenic anion. Since the speciation of arsenic in the investigated groundwater shows that dominant form of arsenic (Table 1) was As(III), which exists in neutral form (as H_3AsO_3) at the pH of the groundwater (pH = 8.05 ± 0.15), the contribution of electrostatic interactions during adsorption of As(III) on the synthesized materials will be insignificant.

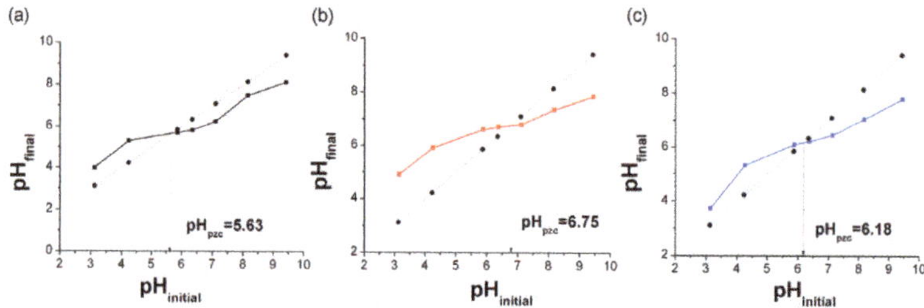

Figure 4. Points of zero charge of the synthesized (**a**) MNp-OP (**b**) MNp-CS and (**c**) MNp-CO.

3.2. Formation of MNp-OP and MNp-CS

The antioxidant capacity of the plant extract is essential for synthesis of the magnetic nanoparticles and is usually in good correlation with phenols contents including flavonoids and tannins [54]. Their concentrations in the onion peel and corn silk extracts were therefore quantified. The results obtained from these tests are given in Table 3.

Table 3. Total phenols, flavonoids and tannins contents of onion peel and corn silk extracts.

	Total Phenols (mg GAE/g dw)	Total Flavonoids (mg QE/g dw)	Total Tannins (mg CE/g dw)	DPPH•IC$_{50}$ (µg/mL)	FRAP (mg TE/g dw)
Onion peel extract	155 ± 4.8	19.9 ± 1.8	133 ± 9.5	8.82 ± 0.87	2.17 ± 0.17
Corn silk extract	756 ± 56.3	205 ± 7.1	532 ± 16	2.93 ± 0.21	6.44 ± 0.32

The corn silk extract has higher phenols, flavonoids and tannins contents than the onion peel extracts. Furthermore, the antioxidant capacity of corn silk extract obtained as DPPH• scavenging and reduction potentials were higher than the onion peel extract, confirming the close relation between the content of phytochemical compounds and the extract's antioxidant properties [55]. Lee et al. [27] investigated different conditions for the extraction of phenolic compounds including flavonoids from onion peel and observed that total phenols and total flavonoids contents for onion peel extracts produced by hot water (80 °C for 3 h) was 120.60 ± 6.05 GAE/g dry weight of extract and 54.50 ± 5.21 mg QE/g dry weight. The total phenols and flavonoids contents in corn silk extract in our study were also found to be comparable with corn silk extracts reported by other researchers [56,57]. It is possible that the mild extract conditions applied in this work (30 °C for 1 h), which were chosen for their significantly lower energy cost, resulted in fewer losses of phenolic compounds due to oxidation [58]. In general, however, the phytochemical content of plants is greatly affected by plant variety, maturity, growing conditions, cultivar areas and harvest times, so it is also possible the relatively high phenol contents in this work are the results of natural variation [35,59].

The characterisation results presented above demonstrate that the addition of $FeCl_3$ to the onion peel or corn silk extracts cause the reduction of Fe^{3+} and subsequent formation of Fe_3O_4 nanoparticles. As can be seen from extract characterization (Table 3), onion shell and corn silk extract contained different phytochemical compounds with high antioxidant activities. In the synthesis of MNp-OP (suggested mechanism shown in Figure 5), we assume that the flavonoid querecetin, the major flavonoid in the onion peel extract, plays a critical role. Initially, carbonyl and hydroxyl groups (3-hydroxo-4-ketopresent) in these molecules chelate with Fe^{3+} to form ferric complex [60]. This metal chelation also has pro-oxidant properties (via electron or H-atom donation), whereas the transfer of an electron to Fe^{3+} resulting in formation of Fe^{2+} where quinones are suggested as flavonoid oxidation products [60]. Finally, Fe^{2+} and excess Fe^{3+} form black precipitate of Fe_3O_4 nanoparticles. Like onion peel, corn silk extract is also rich in phenolic compounds which are capable of reducing Fe^{3+} [61].

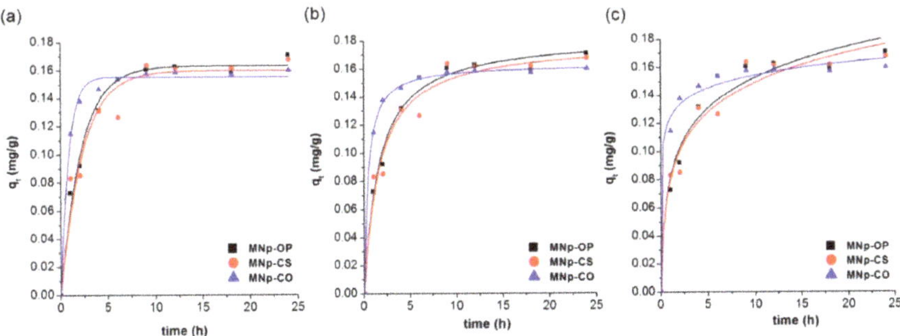

Figure 5. Schematic illustration of preparation of magnetic nanoparticles (MNp-OP) obtained from onion peel extract.

The results above demonstrate that onion peel and corn silk can both be used to produce magnetic nanoparticles, thus representing a significant added value to materials currently disposed of as a non-usable wastes.

3.3. Arsenic Adsorption Kinetics

Kinetic adsorption experiments and their corresponding mathematical models were used to investigate the adsorption rate and mechanism of arsenic removal on the synthesized magnetic nanoparticles. The changes in the adsorption of the synthesized nanoparticles (q_t) of arsenic during time t, are presented in Figure 6.

Figure 6. (**a**) Pseudo-first order (**b**) pseudo-second order and (**c**) Elovich model for adsorption of As on MNp-OP, MNp-CS and MNp-CO.

As can be seen from Figure 6, the adsorption process is time dependent and can be divided into two steps. During the first 4 h, adsorption of arsenic was fast and over this period, arsenic removal was 55–61%, depending on the applied magnetic nanoparticles. In subsequent slower steps, the adsorption of arsenic decreased and reached equilibrium after 9 h in the case of MNp-OP and MNp-CS. In the case of arsenic adsorption on MNp-CO, a contact time of 6 h was sufficient to reach equilibrium.

The adsorption behaviour of arsenic on the synthesized materials can be explained by the basic mechanisms of sorption and the hydrodynamics of the system. Namely, at the beginning of the process, the number of sorption sites and the concentration of arsenic in the solution are maximum, so the driving force of the sorption process is also at a maximum. In addition, mixing provides the energy necessary to transport the arsenic through the liquid film to the active sites on the sorbent. Consequently, sorption is promoted by these three facts: the presence of a large number of active sites, a large driving force and less resistance to mass transfer caused by mixing [62]. Over time, due to the accumulation of arsenic on the sorbent surface and the resulting decrease in the concentration gradient, the rate of arsenic sorption decreases, eventually reaching a state of equilibrium.

In addition, since the equilibrium time in all investigated cases was relatively slow (in the order of hours), this suggests that specific adsorption (which involves the formation of inner surface complexes) of arsenic occurs on these materials. Unlike specific adsorption, adsorption that is achieved only by electrostatic interactions is usually very fast (in the order of seconds) [63].

To obtain more in-depth information about the mechanism of arsenic adsorption onto the synthesized magnetic nanoparticles, the kinetic data was fitted to pseudo-first-order, pseudo-second-order, and Elovich kinetic models [40]. The parameters of these models are presented in Table 4.

Table 4. Kinetic parameters for arsenic adsorption on MNp-OP, MNp-CS and MNp-CO.

		MNp-OP	MNp-CS	MNp-CO
	k_1 (1/h)	0.459	0.443	1.26
Pseudo-first order	q_e (mg/g)	0.164	0.159	0.155
	R^2	0.9867	0.9371	0.9925
	k_2 (g/mg h)	3.43	3.49	15.2
	q_e (mg/g)	0.183	0.179	0.163
Pseudo-second order	q_{eexp} (mg/g)	0.170	0.173	0.177
	h (mg/g h)	0.115	0.112	0.403
	R^2	0.9871	0.9632	0.9990
	α (mg/g h)	0.377	0.382	180
Elovich	β (mg/g)	31.3	32.3	76.6
	R^2	0.9592	0.9607	0.9868

Based on the coefficients of determination (R^2), all applied models show good fit with the experimental data. However, the highest values of R^2 were observed for the pseudo-second-order kinetic model in all cases (Table 4), implying that adsorption process may be controlled by chemisorption [8]. The theoretical q_e values calculated from the pseudo-second order model were very close to the experimental q_e values confirming the good agreement with the model. The initial sorption rate, h, in all cases was higher in comparison to the rate constant k_2, indicating that the rate of adsorption was much faster at the beginning of the process and decreased with time.

In a solid–liquid sorption process, which include porous adsorbents, transfer of adsorbate often takes place over four steps: bulk diffusion, film diffusion (which includes transfer of adsorbate from the bulk liquid phase to the adsorbent's external surface through a hydrodynamic boundary layer or film), intraparticle diffusion (which involves diffusion of adsorbate into the pores of the adsorbent, along pore-wall surfaces, or both), and sorption of sorbate molecules onto active sites distributed within the sorbent particles [64]. The first and last steps are usually very fast and they do not have a determinant role in governing sorption rates. Consequently, the adsorption rate might be controlled by external diffusion, inner diffusion or both.

Since the pseudo-first, the pseudo-second order and Elovich kinetic models cannot identify the influence of diffusion on sorption, the intraparticle diffusion model proposed by Weber and Morris, and the external diffusion model, were also used to fit the experimental data (Table 5, Figures 7 and 8).

Table 5. Intraparticle and external diffusion parameters for arsenic adsorption on MNp-OP, MNp-CS and MNp-CO.

Diffusion Model	Parameters	MNp-OP	MNp-CS	MNp-CO
Weber–Morris model	k_i (mg/g $h^{0.5}$)	0.0576	0.0407	0.0954
	C_i	0.0139	0.0376	0.0250
	R^2	0.9630	0.8844	0.8364
	k_i (mg/g $h^{0.5}$)	0.00510	0.00188	0.150
	C_i	0.143	0.156	0.00195
	R^2	0.5454	0.0319	0.4932
Boyd model	Intercept	0.240	0.254	0.482
	R^2	0.9822	0.9288	0.9942

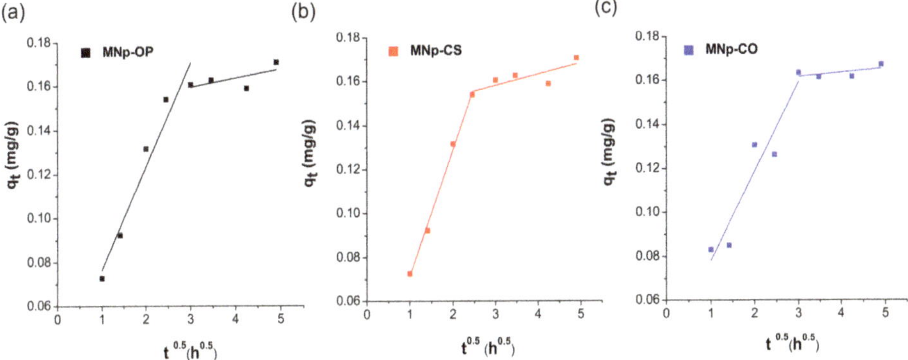

Figure 7. Weber–Morris model for arsenic adsorption on (**a**) MNp-OP (**b**) MNp-CS and (**c**) MNp-CO. Initial As concentration 120 µg/L; adsorbent dose 0.5 g/L; pH 8.2.

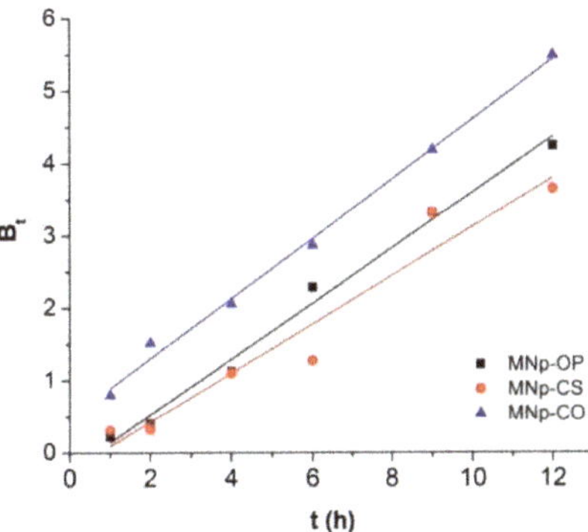

Figure 8. Boyd model for arsenic adsorption on MNp-OP, MNp-CS and MNp-CO. Initial As concentration 120 µg/L; adsorbent dose 0.5 g/L; pH 8.2.

According to the Weber–Morris model, if the plot of q_t vs. $t^{0.5}$ is linear and passes through the origin, adsorption is entirely governed by intraparticle diffusion. In contrast, if the plot is multilinear, it indicates that the adsorption is a multi-stage process controlled by various limiting factors at different steps of the process [8].

As can be seen from Figure 7 all the plots present multi-linearity, imply that intraparticle diffusion was involved in the adsorption process, but not the only rate-controlling step. Moreover, all plots give an intercept C (Table 4) which means that external film mass transfer or boundary layer control also exists.

The contribution of boundary layer or film diffusion is often confirmed using the Boyd model [65]. According to this model, if the plot Bt vs. t, is a straight line passing through the origin, then intraparticle diffusion is the rate controlling step. Otherwise film diffusion determines the process.

From Figure 8, it is observed that all plots t are linear, but the curves do not pass through the origin and have an intercept, suggesting that arsenic sorption on MNp-OP, MNp-CS and MNp-CO was mainly governed by external mass transfer (film diffusion) [66].

In order to make a distinction between kinetic and diffusion control, a very general guideline can be used: if equilibrium is achieved within 3 h, the process is usually kinetic controlled and above 24 h, it is diffusion-controlled [67]. Thus, taking into account that the equilibrium time of arsenic adsorption on MNp-OP, MNp-CS and MNp-CO was 6 h and 9 h, respectively, the good correlation of the experimental data with the pseudo-second order model, and the results obtained by the diffusion model, we suggest that adsorption mechanism on MNp-OP, MNp-CS and MNp-CO is a complex process involving chemical interaction (surface adsorption), external and intraparticle diffusion.

3.4. Arsenic Adsorption Isotherms

Adsorption isotherms are fundamental in describing the interactive behaviour between adsorbate and adsorbent. The adsorption isotherm yields certain constant values, which express the surface properties and affinity of the adsorbent which play an important role in the design of an adsorption system. Different adsorption isotherms have been developed, however in this study, the non-linear form of Langmuir, Freundlich, Temkin and Dubinin–Radushkevich isotherms [41] were used to model the experimental equilibrium data (Figure 9, Table 6).

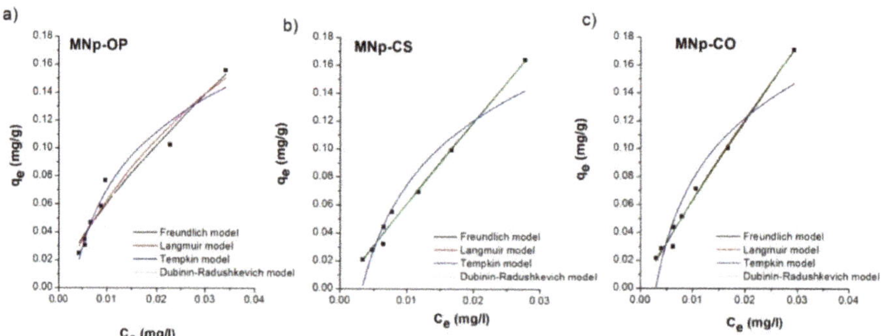

Figure 9. Adsorption isotherms of arsenic on (**a**) MNp-OP (**b**) MNp-CS and (**c**) MNp-CO. (C_0(As) = 120 µg/L; sorbent dose 0.5–5 g/L; pH = 8.1.

Table 6. Parameters of Freundlich, Langmuir, Temkin and Dubinin–Radushkevich sorption isotherm models for As(III) and As(V) sorption on MNp-OP, MNp-CS and MNp-CO.

	MNp-OP	MNp-CS	MNp-CO
Freundlich model			
K_F (mg/g)/(mg/L)^n	4.97	5.22	4.49
n	0.756	0.966	0.925
R^2	0.9620	0.9899	0.9901
Langmuir model			
q_{max} (mg/g)	1.86	2.79	1.30
K_L (L/g)	8.21	2.24	5.15
R_L	0.9524	0.782	0.610
R^2	0.9594	0.9880	0.9881
Dubinin–Radushkevich model			
q_d (mg/g)	4.45	5.64	4.84
k_{ads} (mol^2/kJ^2)	0.000671	0.000397	0.000381
E (kJ/mol)	27.3	35.5	36.2
R^2	0.9473	0.9878	0.9496
Tempkin model			
b_T (kJ/mol)	41.2	37.1	38.6
A_T (L/g)	325	309	341
R^2	0.9252	0.9025	0.8825

The results of the isotherm modelling, based on the coefficient of determination (R^2), revealed that all applied models show good agreement with the experimental data (Table 6). However, the Freundlich model provided slightly better fit for As adsorption on MNp-OP, MNp-CS and MNp-CO (coefficients of determination, R^2, were 0.9620, 0.9899 and 0.9901, respectively), suggesting that multilayer (heterogeneous) adsorption was the preferred removal mechanism [51,68]. This is supported by the SEM images, which showed that the surface of these materials was nonuniform with different particle shapes and sizes.

The Freundlich constant K_f of MNp-CS and MNp-OP was higher than for MNp-CO, implying that the green synthesized adsorbents can be competitive in terms of adsorption affinity with magnetic nanoparticles produced via chemical precipitation method. The parameter, 1/n as a measure of adsorption intensity or adsorbent surface heterogeneity, in all cases was less than 1, indicating that adsorption of As on synthesized magnetic nanoparticles is favourable and chemical in nature [68].

The maximum adsorption capacity obtained from the Langmuir model (q_{max}) of the MNp-CS and MNp-OP was higher than MNp-CO. This trend was in accordance with the K_F values obtained by the Freundlich model. The Langmuir equilibrium constant which reflects the affinity between adsorbent and the adsorbate, K_L (L/g), was 8.21, 2.24 and 5.15 L/g, for MNp-OP, MNp-CS and MNp-CO, respectively, suggesting strong binding of arsenic onto the magnetite nanoparticles. Separation factors (R_L), essential characteristics of the Langmuir model, were in all cases below 1, which means that adsorption of arsenic on synthesized magnetic nanoparticles was favourable [69]. This was consistent with the Freundlich constants 1/n measure of sorption intensity, which also ranged between zero and one.

In the case of the Dubinin–Radushkevich model, R^2 values were in the order of MNp-CS > MNp-CO > MNp-OP. The theoretical sorption capacity (q_d) obtained from this model was the highest for MNp-CS and the lowest for MNp-CO, which could be attributed to its higher specific surface area (Table 2). The free energy of adsorption (E) for MNp-OP, MNp-CS and MNp-CO, were 27.3 kJ/mol, 35.5 kJ/mol and 36.2 kJ/mol, respectively, implying that chemisorption may be the preferred sorption mechanism of As on the magnetic nanoparticles. Namely, when the value of E is less than 20 kJ/mol, the adsorption process is called physisorption, if it is between 20–40 kJ/mol the process is known as ionic exchange, and when E is larger than 40 kJ/mol it is a chemisorption process [70].

The correlation coefficients for the Tempkin isotherm were low in comparison with other applied models ($R^2 = 0.8825–0.9252$). Temkin constants, b_T, which represent the heat of adsorption, were in the range 37.1–41.2 kJ/mol, indicating the exothermic nature of the adsorption process. The value of $b_T > 8$ kJ/mol implied that the interaction between arsenic and synthesized magnetic nanoparticles was fairly strong and was not easily reversible (chemisorption) [71] which is in agreement with the results of the Dubinin–Radushkevich model. The binding constants of magnetic nanoparticles, A_T (L/g), synthesized via green method, was higher than MNp-CO, confirming their higher adsorption capacity as predicted by the Langmuir isotherm.

Note that these experiments were all carried out in real groundwater, which is naturally contaminated with arsenic. It is therefore not possible to directly compare for example the q_{max} values obtained herein with the results of other authors, as a large number of other factors can significantly affect the efficacy of adsorption processes, especially the initial arsenic concentration and the species of arsenic present, the pH of the water and the presence of other anions. For example, phosphate is well known to compete with arsenic in adsorption processes [18] and has a concentration 11 times higher than the arsenic concentration in this groundwater (Table 1). Furthermore, although the arsenic content of the real groundwater investigated is an order of magnitude higher than recommended, it is still an order of magnitude lower than many of the studies carried out in synthetic matrices, where the very high initial arsenic concentrations often result in correspondingly high q_{max} values [16,18,51]. The range of values (1.3 to 2.8 mg/g) obtained in this work compare very favourably with those from our previous work, in which Fe–Mn modified granular activated carbon was investigated with the same groundwater ($q_{max} = 0.5$ mg/g) [72], making these the most effective sorbents we have investigated to date.

Finally, in order to evaluate the stability of synthesized nanoparticles, after the adsorption experiments, the residual iron concentrations in the treated samples were determined (Table 7).

Table 7. Residual Fe concentration in groundwater samples after arsenic adsorption on MNp-OP, MNp-CS and MNp-CO.

Sorbent Dose (g/L)	0.5	1.0	1.5	2.0	2.5	3.5	4.0	5.0
	Residual Fe (mg/L)							
MNp-OP	0.036	0.045	0.062	0.068	0.079	0.116	0.120	0.171
MNp-CS	0.045	0.065	0.089	0.078	0.092	0.134	0.154	0.168
MNp-CO	0.035	0.052	0.065	0.084	0.086	0.120	0.133	0.154

The residual iron concentrations increased with the applied sorbent dose. For high sorbent doses (>2.5 g/L), the maximum allowable concentration (MAC) for iron in drinking water of 0.1 mg/L was exceeded. However, the MAC for arsenic was satisfied at much lower sorbent doses where residual iron concentrations were below the permissible level, suggesting that these adsorbents can be applied safety.

4. Conclusions

Novel magnetic nanoparticles (MNp-OP and MNp-CS) were successfully prepared via a green synthesis approach, using environmentally sound waste materials, onion peel and corn silk extract. The presence and formation of magnetic nanoparticles was confirmed by XRD and FTIR analysis while the high specific surface area and pore volume of these materials suggested they would have good performance for the removal of arsenic from groundwater. In terms of adsorption capacity, the green synthesized magnetic nanoparticles were shown to be highly competitive and effective alternatives to magnetic nanoparticles synthesized by the most common chemical precipitation method (MNp-CO). Adsorption kinetics of arsenic on MNp-OP, MNp-CS, MNp-CO were best described by the pseudo-second-order model, suggesting a chemisorption mechanism, while the Weber–Morris and Boyd model showed that external and intraparticle diffusion contributed to the overall adsorption process. The magnetic properties of these green synthesized adsorbents, which allows for their easy separation from water and which thus eliminates more complex and expensive separation techniques (filtration and centrifugation methods), along with their high potential for arsenic removal from groundwater, makes these materials very promising candidates for water treatment.

Author Contributions: Conceptualization, A.T. and J.A.; Investigation, J.N., M.W., M.Š. and T.M.; Methodology, J.N. and S.M.; Project administration, J.A.; Supervision, J.A.; Visualization, J.N.; Writing—original draft, J.N. and S.M.; Writing—review & editing, A.T. and M.W.

Funding: This research was funded by the Ministry of Education, Science and Technological Development of the Republic of Serbia (Project No. TR37004 and III43005).

Acknowledgments: The authors gratefully acknowledge the support of the Ministry of Education, Science and Technological Development of the Republic of Serbia.

Conflicts of Interest: The authors declare no conflicts of interest.

References

1. Shankar, S.; Shanker, U.; Shikha. Arsenic Contamination of Groundwater: A Review of Sources, Prevalence, Health Risks, and Strategies for Mitigation. *Sci. World J.* **2014**, *2014*, 304524. [CrossRef] [PubMed]

2. Abdul, M.K.S.; Jayasingheb, S.S.; Chandanaa, E.P.S.; Jayasumanac, C.; Mangala, P.; De Silva, C.S. Arsenic and human health effects: A review. *Environ. Toxicol. Pharmacol.* **2015**, *40*, 828–846. [CrossRef] [PubMed]

3. World Health Organisation (WHO). *Guidelines Drinking-Water Quality*, 4th ed.; WHO: Geneva, Switzerland, 2011.

4. Nicomel, N.N.; Leus, K.; Folens, K.; De Voort, P.V.; Laing, G.D. Technologies for Arsenic Removal from Water: Current Status and Future Perspectives. *Int. J. Environ. Res. Public Health* **2016**, *13*, 62. [CrossRef] [PubMed]

5. Pathan, S.; Bose, S. Arsenic Removal Using "Green" Renewable Feedstock-Based Hydrogels: Current and Future Perspectives. *ACS Omega* **2018**, *3*, 5910–5917. [CrossRef]

6. Mukherjeea, D.; Ghosha, S.; Majumdar, S.; Annapurnaa, K. Green synthesis of α-Fe$_2$O$_3$ nanoparticles for arsenic(V) remediation with a novel aspect for sludge management. *J. Environ. Chem. Eng.* **2016**, *4*, 639–650.

7. Poguberović, S.; Krčmar, D.; Maletić, S.; Kónya, Z.; Tomašević Pilipović, D.; Kerkez, Đ.; Rončević, S. Removal of As(III) and Cr(VI) from aqueous solutions using "green" zero-valent iron nanoparticles produced by oak, mulberry and cherryleaf extracts. Removal of As(III) and Cr(VI) from aqueous solutions using "green" zero-valent iron nanoparticles produced by oak, mulberry and cherryleaf extracts. *Ecol. Eng.* **2016**, *90*, 42–49.

8. Medina-Ramirez, A.; Gamero-Melo, P.; Ruiz-Camacho, B.; Minchaca-Mojica, J.I.; Romero-Toledo, R.; Gamero-Vega, K.Y. Adsorption of Aqueous As(III) in Presence of Coexisting Ions by a Green Fe-Modified W Zeolite. *Water* **2019**, *11*, 281. [CrossRef]

9. Asere, T.G.; Stevens, C.V.; Laing, G.D. Use of (modified) natural adsorbents for arsenic remediation: A review. *Sci. Total Environ.* **2019**, *676*, 706–720. [CrossRef]

10. Habuda-Stanić, M.; Nujić, M. Arsenic removal by nanoparticles: A review. *Environ. Sci. Pollut. Res.* **2015**, *22*, 8094–8123. [CrossRef]

11. Siddiqui, S.I.; Chaudhry, S.A. Arsenic removal from water using nano-composites: A review. *Curr. Environ. Eng.* **2017**, *4*, 81–102. [CrossRef]

12. Wong, W.W.; Wong, H.Y.; Borhan, A.; Badruzzaman, M.; Goh, H.H.; Zaman, M. Recent advances in exploitation of nanomaterial for arsenic removal from water: A review. *Nanotechnology* **2017**, *28*, 1–31. [CrossRef] [PubMed]

13. Song, K.; Kim, W.; Suh, C.Y.; Shin, D.; Ko, K.S.; Ha, K. Magnetic iron oxide nanoparticles prepared by electrical wire explosion for arsenic removal. *Powder Technol.* **2013**, *246*, 572–574. [CrossRef]

14. Lunge, S.; Singh, S.; Sinha, A. Magnetic iron oxide (Fe$_3$O$_4$) nanoparticles from tea waste for arsenic removal. *J. Magn. Magn. Mater.* **2014**, *356*, 21–31. [CrossRef]

15. Roya, P.K.; Choudhury, M.R.; Ali, M.A. As (III) and As (V) Adsorption on Magnetite Nanoparticles: Adsorption Isotherms, Effect of pH and phosphate, and Adsorption Kinetics. *Int. J. Chem. Environ. Eng.* **2013**, *4*, 55–63.

16. Tuutijärvi, T.; Lu, J.; Sillanpää, M.; Chen, G. As(V) adsorption on maghemite nanoparticles. *J. Hazard. Mater.* **2009**, *166*, 1415–1420. [CrossRef]

17. Haw, C.Y.; Mohamed, F.; Chia, C.H.; Radiman, S.; Zakaria, S.; Huang, N.M.; Lim, H.N. Hydrothermal synthesis of magnetite nanoparticles as MRI contrast agents. *J. Ceram. Int.* **2010**, *36*, 1417–1422. [CrossRef]

18. Lin, S.; Lu, D.; Liu, Z. Removal of arsenic contaminants with magnetic-Fe_2O_3 nanoparticles. *Chem. Eng. J.* **2012**, *211*, 46–52. [CrossRef]

19. Ebrahiminezhad, A.; Zare-Hoseinabadi, A.; Sarmah, A.K.; Taghizadeh, S.; Ghasemi, Y.; Berenjian, A. Plant-Mediated Synthesis and Applications of Iron Nanoparticles. *Mol. Biotechnol.* **2018**, *60*, 154–168. [CrossRef]

20. Latha, N.; Gowri, M. Bio synthesis and characterization of Fe_3O_4 nanoparticles using Caricaya Papaya leaves extract. *Synthesis* **2014**, *3*, 1551–1556.

21. Martínez-Cabanas, M.; López-García, M.; Barriada, J.L.; Herrero, R.; de Vicente, M.E.S. Green synthesis of iron oxide nanoparticles. Development of magnetic hybrid materials for efficient As (V) removal. *Chem. Eng. J.* **2016**, *301*, 83–91. [CrossRef]

22. Awwad, A.M.; Salem, N.M. A green and facile approach for synthesis of magnetite nanoparticles. *J. Nanosci. Nanotechnol.* **2012**, *2*, 208–213. [CrossRef]

23. Kumar, B.; Smita, K.; Cumbal, L.; Debut, A. Biogenic synthesis of iron oxide nanoparticles for 2-arylbenzimidazole fabrication. *J. Saudi Chem. Soc.* **2014**, *18*, 364–369. [CrossRef]

24. Maheswari, K.C.; Sreenivasula Reddy, P. Green Synthesis of Magnetite Nanoparticles through Leaf Extract of *Azadirachta indica*. *J. Nanosci. Technol.* **2016**, *2*, 189–191.

25. Venkateswarlu, S.; Yoon, M. Rapid removal of cadmium ions using green synthesized Fe_3O_4 nanoparticles capped with diethyl-4-(4 amino-5-mercapto-4H-1,2,4-triazol-3-yl) phenyl phosphonate. *RSC Adv.* **2015**, *5*, 65444–65453. [CrossRef]

26. Sharma, K.; Mahato, N.; Nile, S.H.; Lee, E.T.; Lee, Y.R.; Mahato, N.; Nile, S.H.; Lee, E.T.; Lee, Y.R. Economical and environmentally-friendly approaches for usage of onion (*Allium cepa* L.) waste. *Food Funct.* **2016**, *7*, 3354–3369. [CrossRef] [PubMed]

27. Lee, K.A.; Kim, K.T.; Kim, H.J.; Chung, M.S.; Chang, P.S.; Park, H.; Paik, H.D. Antioxidant Activities of Onion (*Allium cepa* L.) Peel Extracts Produced by Ethanol, Hot Water, and Subcritical Water Extraction. *Food Sci. Biotechnol.* **2014**, *23*, 615–621. [CrossRef]

28. Bonaccorsi, P.; Caristi, C.; Gargiulli, C.; Leuzzi, U. Flavonol glucosides in *Allium* species: A comparative study by means of HPLC–DAD–ESI-MS–MS. *Food Chem.* **2008**, *107*, 1668–1673. [CrossRef]

29. Benítez, V.; Mollá, E.; Martín-Cabrejas, M.A.; López-Andréu, J.F.; Downes, K.; Terry, L.A.; Esteban, R.M. Study of bioactive compound content in different onion sections. *Plant Foods Hum. Nutr.* **2011**, *66*, 48–57. [CrossRef]

30. Ifesan, B.O.T. Chemical Composition of Onion Peel (*Allium cepa*) and its Ability to Serve as a Preservative in Cooked Beef. *Int. J. Sci. Res. Methodol.* **2017**, *7*, 25–34.

31. Aukkanita, N.; Kemngoena, T.; Ponharna, N. Utilization of Corn Silk in Low Fat Meatballs and Its Characteristics. *Procedia-Soc. Behav. Sci.* **2015**, *197*, 1403–1410. [CrossRef]

32. Nurhanan, A.R.; Rosli, W.I.W.; Mohsin, S.S.J. Total polyphenol content and Free radical scavenging activity of Cornsilk (*Zea mays* hairs). *Sains Malays.* **2012**, *41*, 1217–1221.

33. Nurhanan, A.R.; Rosli, W.I.W. Nutritional compositions and antioxidative capacity of the silk obtained from immature and mature corn. *J. King Saud Univ. Sci.* **2014**, *26*, 119–127.

34. Fatima, A.; Agrawal, P.; Singh, P.P. Herbal option for diabetes: An overview. *Asian Pac. J. Trop. Biomed.* **2012**, *2*, S536–S544. [CrossRef]

35. Sarepoua, E.; Tangwongchai, R.; Suriharn, B.; Kamol, L. Influence of variety and harvest maturity on phytochemical content corn silk. *Food Chem.* **2015**, *169*, 424–429. [CrossRef] [PubMed]

36. Venkateswarlu, S.; Subba Rao, Y.; Balaji, T.; Prathima, B.; Jyothi, N.V.V. Biogenic synthesis of Fe_3O_4 magnetic nanoparticles using plantain peel extract. *Mater. Lett.* **2013**, *100*, 241–244. [CrossRef]

37. Beara, I.N.; Lesjak, M.M.; Jovin, E.D.; Balog, K.J.; Anačkov, G.T.; Orčić, D.Z.; Mimica-Dukić, N.M. Plantain (*Plantago* L.) species as novel sources of flavonoid antioxidants. *J. Agric. Food Chem.* **2009**, *57*, 9268–9273. [CrossRef] [PubMed]

38. Beara, I.N.; Torović, L.D.; Pintać, D.; Majkić, T.M.; Orčić, D.Z.; Mimica-Dukić, N.M.; Lesjak, M.M. Polyphenolic profile, antioxidant and neuroprotective potency of grape juices and wines from Fruška Gora region (Serbia). *Int. J. Food Prop.* **2018**, *20*, S2552–S2568. [CrossRef]

39. Zhang, G.; Liu, H.; Qu, J.; Jefferson, W. Arsenate uptake and arsenite simultaneous sorption and oxidation by Fe–Mn binary oxides: Influence of Mn/Fe ratio. pH. Ca^{2+} and humic acid. *J. Colloid. Interface Sci.* **2012**, *366*, 141–146. [CrossRef]

40. Largitte, L.; Pasquier, R. A review of the kinetics adsorption models and their application to the adsorption of lead by an activated carbon. *Chem. Eng. Res. Des.* **2016**, *109*, 495–504. [CrossRef]

41. Foo, K.Y.; Hameed, B.H. Insights into the modeling of adsorption isotherm systems. *Chem. Eng. J.* **2010**, *156*, 2–10. [CrossRef]

42. USEPA. *Method 200.8: Determination of Trace Elements in Waters and Wastes by Inductively Coupled Plasma-Mass Spectrometry*; Revision 5.4; USEPA: Cincinnati, OH, USA, 1994.

43. *Arsenic Speciation in Urine Becomes Routine Using Agilent HPLC with 7500 Series ICP-MS*; 5989-8399EN; Agilent Technologies: Santa Clara, CA, USA, 2008.

44. AWWA-APHA-WEF. *Standard Methods for the Examination of Water and Wastewater*, 20th ed.; American Public Health Association/American Water Works Association/Water Environment Federation: Washington, DC, USA, 2012.

45. *Water Quality—Determination of Phosphorus—Ammonium Molybdate Spectrometric Method*; SRPS EN ISO 6878:2008; Institute for Standardization of Serbia: Belgrade, Serbia, 2008.

46. *Water Quality—Determination of Chloride—Silver Nitrate Titration with Chromate Indicator (Mors Method)*; SRPS ISO 9297/1:2007; Institute for Standardization of Serbia: Belgrade, Serbia, 2007.

47. *Water Quality—Determination of Ammonium—Distillation and Titration Method*; SRPS ISO 5664:1992; Institute for Standardization of Serbia: Belgrade, Serbia, 1992.

48. *Guidelines for Determination of Total Organic Carbon (TOC) and Dissolved Organic Carbon (DOC) in Water*; SRPS ISO 8245:2007; Institute for Standardization of Serbia: Belgrade Serbia, 2007.

49. Mahdavi, M.; Namvar, F.; Ahmad, B.; Rosfarizan, M. Green Biosynthesis and Characterization of Magnetic Iron Oxide (Fe_3O_4) Nanoparticles Using Seaweed (*Sargassum muticum*) Aqueous Extract. *Molecules* **2013**, *18*, 5954–5964. [CrossRef] [PubMed]

50. Scherrer, P. Bestimmung der Grosse und der Inneren Struktur von Kolloidteilchen Mittels Rontgenstrahlen, Nachrichten von der Gesellschaft der Wissenschaften, Gottingen. *Math.-Phys. Kl.* **1918**, *2*, 98–100.

51. Wen, Z.; Dai, C.; Zhu, Y.; Zhang, Y. Synthesis of ordered mesoporous iron manganese bimetal oxides for arsenic removal from aqueous solutions. *RSC Adv.* **2015**, *5*, 4058–4068. [CrossRef]

52. Kong, S.; Wang, Y.; Zhan, H.; Yuan, S.; Yu, M.; Liu, M. Adsorption/Oxidation of Arsenic in Groundwater by Nanoscale Fe-Mn Binary Oxides Loaded on Zeolite. *Water Environ. Res.* **2014**, *86*, 147–155. [CrossRef] [PubMed]

53. Sing, K.S.W.; Evertt, D.H.; Haul, R.A.W.; Moscou, L.R.A.; Pierotti, J.; Rouquerol, T.; Siemieniewska, T. Reporting Physisorption Data for Gas/Solid Systems with Special Reference to the Determination of Surface Area and Porosity. *Pure Appl. Chem.* **1985**, *57*, 603–619. [CrossRef]

54. Ignat, I.; Volf, I.; Popa, V.I. A critical review of methods for characterisation of polyphenolic compounds in fruits and vegetables. *Food Chem.* **2011**, *126*, 1821–3185. [CrossRef]

55. Machado, S.; Pinto, S.L.; Grosso, J.P.; Nouws, H.P.A.; Albergaria, J.T.; Delerue-Matos, C. Green production of zero-valent iron nanoparticles using tree leaf extracts. *Sci. Total Environ.* **2013**, *445–446*, 1–8. [CrossRef]

56. Nurhanan, A.R.; WR, W.I. Evaluation of polyphenol content and antioxidant activities of some selected organic and aqueous extracts of cornsilk (*Zea mays* hairs). *J. Med Bioeng.* **2012**, *1*, 48–51.

57. Irawaty, W.; Ayucitra, A.; Indraswati, N. Radical scavenging activity of various extracts and varieties of corn silk. *ARPN J. Eng. Appl. Sci.* **2018**, *13*, 10–16.

58. Viera, V.B.; Piovesan, N.; Rodrigues, J.B.; Mello, R.O.; Prestes, R.C.; Santos, R.C.V.; Vaucher, R.A.; Hautrive, T.P.; Kubota, E.H. Extraction of phenolic compounds and evaluation of the antioxidant and antimicrobial capacity of red onion skin (*Allium cepa* L.). *Int. Food Res. J.* **2017**, *24*, 990–999.

59. El Amrani, F.B.; Perello, L.; Real, J.A.; Gonzalez-Alvarez, M.; Alzuet, G.; Borras, J.; Garcia-Granda, S.; Montejo-Bernardo, J. Oxidative DNA cleavage induced by an iron(III) flavonoid complex: Synthesis, crystal structure and characterization of chlorobis (flavonolato)(methanol) iron(III) complex. *J. Inorg. Biochem.* **2006**, *100*, 1208–1218. [CrossRef] [PubMed]

60. Xu, G.R.; In, M.Y.; Yuan, Y.; Joon Lee, J.; Kim, S. In situ Spectroelectrochemical Study of Quercetin Oxidation and Complexation with Metal Ions in Acidic Solutions. *Bull. Korean Chem. Soc.* **2007**, *28*, 889–892.

61. Ebrahimzadeh, M.A.; Pourmorad, F.; Bekhradnia, A.R. Iron chelating activity, phenol and flavonoid content of some medicinal plants from Iran. *Afr. J. Biotechnol.* **2008**, *7*, 3188–3192.

62. Mondal, P.; Mohanty, B.; Majumder, C.B. Removal of Arsenic from Simulated Groundwater by GAC-Fe: A Modeling Approach. *AIChE J.* **2009**, *55*, 1860–1871. [CrossRef]

63. Zhang, G.; Liu, H.; Liu, R.; Qu, J. Adsorption behavior and mechanism of arsenate at Fe–Mn binary oxide/water interface. *J. Hazard. Mater.* **2009**, *168*, 820–825. [CrossRef]

64. Tran, H.N.; You, S.J.; Bandegharaei, A.H.; Chao, H.P. Mistakes and inconsistencies regarding adsorption of contaminants from aqueous solutions: A critical review. *Water Res.* **2017**, *120*, 88–116. [CrossRef]

65. Boyd, G.E.; Schubert, J.; Adamson, A.W. The exchange adsorption of ions from aqueous solutions by organic zeolites. Ion-exchange equilibria. *J. Am. Chem. Soc.* **1947**, *69*, 2818–2829. [CrossRef]

66. Kajjumba, S.W.; Emik, S.; Öngen, A.; Özcan, H.K.; Aydın, S. Modelling of Adsorption Kinetic Processes—Errors, Theory and Application. In *Advanced Sorption Process Applications*; Edebali, S., Ed.; IntechOpen: Rijeka, Croatia, 2018.

67. Ho, Y.S.; Ng, J.C.Y.; McKay, G. Kinetics of pollutant sorption by biosorbents: Review. *Sep. Purif. Method* **2000**, *29*, 189–232. [CrossRef]

68. Niazi, N.K.; Bibi, I.; Shahid, M.; Ok, Y.S.; Burton, E.D.; Wang, H.; Shaheen, S.M.; Rinklebe, J.; Lüttge, A. Arsenic removal by perilla leaf biochar in aqueous solutions and groundwater: An integrated spectroscopic and microscopic examination. *Environ. Pollut.* **2018**, *232*, 31–41. [CrossRef]

69. Mudzielwana, R.; Gitari, M.W.; Ndungu, P. Uptake of As(V) from Groundwater Using Fe-Mn Oxides Modified Kaolin Clay: Physicochemical Characterization and Adsorption Data Modeling. *Water* **2019**, *11*, 1245. [CrossRef]

70. Monárrez-Corderoa, B.E.; Amézaga-Madrida, P.; Leyva-Porrasa, C.C.; Pizá-Ruiza, P.; Miki-Yoshidaa, M. Study of the Adsorption of Arsenic (III and V) by Magnetite Nanoparticles Synthesized via AACVD. *Mater. Res.* **2016**, *19*, 1–10.

71. Shafique, U.; Ijaz, A.; Salman, M.; Zaman, W.; Jamil, N.; Rehman, R.; Javaid, A. Removal of arsenic from water using pine leaves. *J. Taiwan Inst. Chem. Eng.* **2012**, *43*, 256–263. [CrossRef]

72. Nikić, J.; Agbaba, J.; Watson, M.A.; Tubić, A.; Šolić, M.; Maletić, S.; Dalmacija, B. Arsenic adsorption on Fe–Mn modified granular activated carbon (GAC–FeMn): Batch and fixed-bed column studies. *J. Environ. Sci. Health Part A* **2019**, *54*, 168–178. [CrossRef] [PubMed]

Article

Phosphate Induced Arsenic Mobilization as a Potentially Effective In-Situ Remediation Technique—Preliminary Column Tests

Martin V. Maier [1,*], Yvonne Wolter [1], Daniel Zentler [1], Christian Scholz [1], Charlotte N. Stirn [2] and Margot Isenbeck-Schröter [1]

[1] Institute of Earth Sciences, Ruprecht-Karls-University Heidelberg, 69120 Heidelberg, Germany
[2] Institute of Geography, Ruprecht-Karls-University Heidelberg, 69120 Heidelberg, Germany
* Correspondence: martin.maier@geow.uni-heidelberg.de; Tel.: +49-6221-546-004

Received: 28 September 2019; Accepted: 5 November 2019; Published: 12 November 2019

Abstract: Arsenic (As) contamination of groundwater is commonly remediated by pump and treat. However, this technique is difficult to apply or maintain efficiently because the mobility of arsenic varies depending on the geochemical aquifer conditions. Arsenic interacting with the sediment can cause strong retardation, which is counteracted by ions competing for sedimentary sorption sites like silica, bicarbonate and phosphate. Phosphate competes most effectively with arsenic for sorption sites due to its chemical similarity. To accelerate an ongoing but ineffective pump and treat remediation, we examined the competitive effect of increasing phosphate doses on contaminated aquifer material of different depths and thus under distinct geochemical conditions. In the columns with phosphate addition, significant amounts of arsenic were released rapidly under oxic and anoxic conditions. In all tests, the grade of leaching was higher under anoxic conditions than under oxic conditions. As(III) was the dominant species, in particular during the first release peaks and the anoxic tests. Higher amounts of phosphate did not trigger the arsenic release further and led to a shift of arsenic species. We suggest that the competitive surface complexation is the major process of arsenic release especially when higher amounts of phosphate are used. Commonly arsenic release is described at iron reducing conditions. In contrast, we observed that a change in prevailing redox potential towards manganese reducing conditions in the oxic tests and iron reducing conditions in the anoxic column took place later and thus independently of arsenic release. The reduction of As(V) to As(III) under both redox conditions is presumed to be an effect of microbial detoxification. A loss of sulphate in all columns with phosphate indicates an increased microbial activity, which might play a significant role in the process of arsenic release. Preliminary tests with sediment material from a contaminated site showed that phosphate additions did not change the pH value significantly. Therefore, a release of other metals is not likely. Our results indicate that in-situ application of phosphate amendments to arsenic-contaminated sites could accelerate and enhance arsenic mobility to improve the efficiency of pump and treat remediation without negative side effects. The novelty of this approach is the use of only small amounts of phosphate in order to stimulate microbial activity in addition to surface complexation. Therefore, this method might become an innovative and cost-effective remediation for arsenic contaminated sites.

Keywords: arsenic; phosphate; competitive surface complexes; release; mobility; remediation

---

1. Introduction

Remediating arsenic (As) contamination is challenging and generally pump and treat technology is applied. For this method, water is extracted from the contaminated aquifer and treated on-site to reduce the contamination from the source area and to avoid the contaminant from spreading. Iron(II)

sulphate and iron(II) chloride are commonly used as reactants to treat the extracted water (e.g., [1,2]). The effectiveness of this approach is often limited due to the geochemistry of the remediation site. In addition, a continuous input of energy is required. If the treatment is inefficient or even terminated because of decreasing efficiency, the contaminant remains.

Aquifer properties like redox conditions and mineral composition are the main factors controlling arsenic reactions and thus mobility, which can differ considerably from site to site [3,4]. Under neutral conditions there are two redox species of inorganic arsenic—Arsenite (H_3AsO_3), where arsenic is in the trivalent oxidation state As(III) and Arsenate ($HAsO_4^{2-}$ and $H_2AsO_4^{-}$) in the pentavalent oxidation state As(V). Under oxic conditions As(V) is the dominant redox state. As(V) is less mobile than As(III) because it forms complexes on mineral surfaces, especially on hydroxides and oxides of iron(III) [5–7] as well as on calcite [8]. Under iron reducing conditions arsenic is released from those surfaces into groundwater and remains mobile after subsequent reduction to As(III) [4,9,10]. In contrast to this, arsenic precipitates as sulphide or forms sulphide complexes if sulphate-reducing conditions prevail and sulphate is available [11,12].

Instead of water extraction, more recent remediation strategies aim to mobilize or immobilize arsenic within the aquifer [13]. In remediation practice, different amendments are used in-situ to decrease arsenic mobility by creating a suitable redox milieu for the sorption of arsenic on iron minerals or the precipitation of iron sulphide minerals [14–20]. However, through immobilization, the contaminant remains in the subsoil and a remobilization cannot be ruled out. A different approach is to increase arsenic mobility by implementing iron reducing conditions in combination with a hydraulic barrier (e.g., pump and treat) to remove the contaminant [21,22]. Artificially implemented iron reducing conditions (e.g., with lactate) are difficult to maintain stable during the entire remediation process and undesired impacts on the aquifer's geochemistry, such as heavy metal release, are a possible side effect.

Phosphate is an opponent substance for arsenate in building up strong surface complexes on iron(hydr-)oxides [23–25]. In pH neutral conditions, the affinity of phosphate to form surface complexes on iron minerals like goethite and gibbsite phases is higher than the one of As(V) [26,27]. Recent studies indicate similar processes with calcite, although the sorption of arsenic on calcite is less favoured than onto iron minerals [28]. It may become a relevant reaction when the proportion of iron minerals in the aquifer sediment is low or if hydrogen carbonate dominates. The reaction kinetics are discussed controversially. Some studies assume an incorporation of arsenic into calcite [29], while others show that arsenic is bound in surface complexation [8,30]. Results of both studies agree that predominantly the As(V)-species interacts with calcite, which can be replaced by phosphate.

Arsenic release from sediment by phosphate addition has been topic to research since the early 70s and recent studies show that the high complexity of the process limits the predictability for field application [31–33]. One study found that high phosphate dosages up to 300 mM to sediment samples under oxic conditions led to arsenic release rates of 80% [33]. Neither longer reaction nor repeated treatment could significantly increase the arsenic leaching. A different study demonstrated that pulse injections of phosphate could increase the total desorption of arsenic up to 94% [34]. Even low phosphate amendments of 6 mM could rapidly desorb more than 35% of the total arsenic from the sediment and continuing amendments led to up to 65% desorption [35].

The novelty of the approach presented here is an increase of arsenic mobility without changing the aquifer's geochemistry for future field application [36,37]. Phosphate is known to compete with arsenate for sedimentary sorption sites but so far there is no information about the effect of different amounts of phosphate on the mobility of arsenic under different redox conditions. In order to investigate the driving processes, we conducted column tests with amendment of varying phosphate concentrations under oxic and anoxic conditions and closely monitored arsenic species distribution in the sediment and the column outflow. In contrast to previous studies [34,35], low phosphate concentrations were used to identify the relevant mechanisms underlying arsenic mobility and speciation. This preliminary laboratory study is the basis for further research to develop an innovative and more efficient remediation field method.

2. Materials and Methods

2.1. Aquifer Material, Groundwater Sampling and Analysis

The aquifer sediment material was collected during previous investigations at the contaminated site in Lampertheim, Germany [36]. Sediments were gathered through percussion drilling with polyethylene-lined probes (d = 100 mm; l = 1 m) to the aquifer basis (30 m b.g.l.). The liners were sealed immediately after recovery to avoid changes of the predominant redox milieu in the aquifer. The cores were opened lengthwise under argon atmosphere for sediment profile description and sampling for geochemical analyses. The cores were covered with plastic wrap and stored at −18 °C. For the column tests, these cores were defrosted and prepared under argon atmosphere. Sediments from the partially water-saturated (5 to 6 m) and completely water-saturated (11 to 12 m) depths of the aquifer were used to ensure they contain the redox-adapted microfauna.

Sedimentary characteristics like grain size, moisture content, smell and colour were determined according to standard regulations (ISO 14688). For the analytical procedures and the column preparation the sediment was homogenized under argon atmosphere. Analytical procedures were performed in replicate to ensure reproducibility. Chemicals used were of per analysis (p.a.) quality. Aqua regia extractions were done once before and once after the column tests and with the sediment standard BAM U112a. Therefore, 1 g of moist sample was mixed with 3 mL of nitric acid (65% HNO_3 p.a., J.T. Baker) and 9 mL of hydrochloric acid (37% HCl p.a., J.T. Baker), covered with watch glasses and simmered for about 3 h at about 130 °C (method modified after ISO11466:1995). After cooling, the samples were filled up to 50 mL with distilled water, filtered through a pleated paper filter and stored at 4 °C until measurement.

In order to distinguish how the different arsenic species are bound to the sediment surfaces, fractionated elutions—based on the method by Reference [38]—were made before and after the column tests. Under argon atmosphere 3.5 g of moist sediment was weighed into four vessels respectively. Distilled water (40 mL) was added to two of the samples and 40 mL of 1 M sodium dihydrogen phosphate ($NaH_2PO_4 \cdot H_2O$ p.a., Grüssing) to the other two. The closed samples were stirred for 24 h on a shaking table, centrifuged (5 min at 4000 rpm) and then filtrated through 0.45 μm membrane filters. The samples were acidified with 6 M hydrochloric acid to pH 1 and stored at 4 °C until measurement.

Stratified groundwater samples from the study site were taken based on the regulations of ISO 5667-11:2009 with a Grundfos MP1 (Grundfos GmbH, Erkrath, Germany) and a packer system after 30 min of pumping. Temperature (SenTix® 940, WTW, Weilheim, Germany), oxygen (CellOx® 325, WTW, Weilheim, Germany), electrical conductivity (Tetracon® 925, WTW, Weilheim, Germany), pH value (SenTix® 940, WTW, Weilheim, Germany) and redox potential (SenTix® ORP-T 900, WTW, Weilheim, Germany), were measured in a flow cell with a WTW Multiline (Multi 3420, WTW, Weilheim, Germany). Results are presented in Table A1. Samples for cations, anions and carbon species were filtered with 0.45 μm syringe membrane filters and cation samples stabilized with HNO_3, arsenic samples with 6 M HCl (p.a. Fluka) to pH 1.

Water cation samples (groundwater and column outflow) and the extractions/elutions from the sediment samples were measured with an ICP-OES (VISTA-MPX CCD Simultaneous, Varian/Agilent, Mulgrave, Australia) for As, Fe, Mn, Ca, Mg, K and Na after standard-procedures and calibrated with element standards (Ultra Scientific), the ICP multi-element standard solution IV (Merck) and validated with two reference materials (SPW-SW2 and TMDA-51.3). The calculated error of the measurement was between 1 and 15%. Arsenic species were measured with a flow-injection hybrid generating system (FI-HG) coupled to the ICP-OES according to Reference [39]. Species were determined in two measurement steps as described in [40] and with a subsequent iterative calculation of arsenic. Water anion samples were analysed with ion chromatography (Dionex DX-120, ThermoFisher Scientific, Sunnyvale, CA, USA) for Cl, Br, NO_3, SO_4 and F with an AS9 HC column and standard calibration for the measured elements (Merck). Total carbon (TC) and inorganic carbon (IC) of the groundwater were measured with a TOC-V$_{CPH}$ (Shimadzu, Kyoto, Japan), from which the total organic carbon (TOC) content was calculated. A total inorganic carbon standard (Ultra Scientific) was used for calibration. Phosphate concentrations from the column outflow were determined with the molybdenum blue method and measured at 880 nm with a photometer (Specord 50, Analytik Jena, Jena, Germany). Total carbon of sediment samples was determined with a sulphur/carbon analyser (Leco SC 144DR, San Joseph, MI, USA). Measurement limits, used reference materials and standard deviations are given in Table 2.

The geochemical conditions in the columns were deduced from hydrogeochemical water composition with respect to redox-sensitive and pH-sensitive parameters like iron, manganese and bicarbonate (Table A1). Because the aquifer material at the study site is carbonate-dominated, the system is sufficiently buffered and measurements of pH values at the column outflow were hence omitted. Redox conditions can be inferred more accurately from the prevailing geochemistry than from redox measurements. Overall measurement quality and methodological errors were evaluated through calibrations and standards, which are presented in Table 2. Ion balance calculations after [41] were within acceptable limits (10% deviation) for the samples without phosphate addition.

2.2. Column Tests

All columns were designed identically with exception of the column diameters. The columns were filled under argon atmosphere with the sandy aquifer material from the contaminated site. For the oxic columns, sediments of 5 to 6 m depth were used, which was within the groundwater fluctuation zone and represents oxic to suboxic conditions. For the anoxic columns, sediments from below the groundwater table in 11 to 12 m depth were taken, representing anoxic conditions (Table 3). Glass wool at both ends of the columns allowed an even flow inside the columns. Top and bottom were sealed airtight with a screw top containing an outlet pipe. The columns were oriented vertically and connected to a peristaltic pump (Minipuls 2, Gilson, Middleton, WI, USA) to obtain a constant upward oriented flow (Figure 1). Through sodium bromide tracer tests the hydrodynamic properties were determined in each column after the leaching tests. Therefore, a 20 mg/L concentration of tracer was added continuously to the test water for 44 h. Samples were taken every 30 min and measured with IC (Dionex DX-120, ThermoFischer Scientific, Sunnyvale, CA, USA) and a WTW multiline (bromide probe Br 500 and electrical conductivity, both WTW, Weilheim, Germany) respectively. The hydrodynamic dispersivities and effective porosities were calculated from breakthrough curves for each column using a one-dimensional analytical solution of the transport equation after Kinzelbach (1987), which was implemented in an Excel worksheet model [42,43].

Chemical analyses of the groundwater at the research site showed predominantly earth alkaline to alkaline water with high contents of hydrogen carbonate. For the column tests replicated groundwater with a composition according to the groundwater in the corresponding depth of the aquifer was prepared every four days and the ingredients (Table 4) were dissolved by carbon dioxide addition with a soda machine. For the anoxic columns, the process-water was degassed with argon for four days before application in order to remove oxygen.

The flow discharge varied from 180 to 220 mL/day respectively, depending on the column diameter. Flow rates were adapted correspondingly to 1 and 1.2 pore volumes (PV) per day to ensure comparable flushing of both columns. Accordingly, the effective porosity of the column fillings was 25 to 30% as calculated from the tracer test.

The experimental conditions of each column are shown in Table 5. All tests were initially run with replicated groundwater without phosphate for eight pore volumes. The anoxic columns were operated strictly anoxic in an argon-filled glove box. Tests with phosphate amendment were done in duplicate (columns 2, 3 and 5, 6, results of duplicates shown in Figures 3 and 4) to consider random errors. Reference columns (column 1 and 4) were run without phosphate amendments.

The timing of phosphate increments depended on the observed arsenic release. Deviations from the scheduled values were caused by phosphate precipitation in the storage vessels. Influent and effluent water was sampled daily to determine the actual concentration of the reactant and to calculate balances.

A fraction collector was used for sampling of columns 1, 2, 4 and 5 and a collective daily sample was taken for columns 3 and 6. The applied analytical methods and sample preparation are listed in Table 6 and described in Section 2.1.

3. Results

3.1. Oxic Tests

In the oxic tests without phosphate addition, (column 1, Figure 1a) the initial concentration of arsenic was about 300 µg/L and decreased quickly to level off at approximately 50 µg/L. At the beginning of the test, As(III) was the dominating species. After 20 PV the shares of As(V) gradually increased while As(III) decreased. By the end of the experiment (PV 44), As(V) was the only remaining species. In total 0.67 mg of arsenic was released from the column.

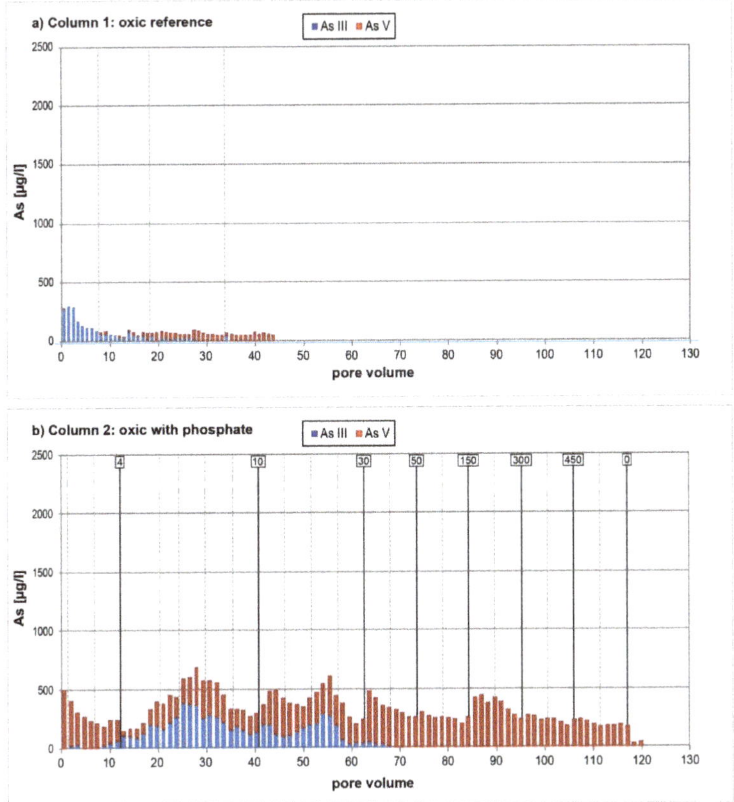

Figure 1. Arsenic release during the oxic column test, (**a**) column 1 without phosphate addition and (**b**) column 2 with phosphate addition (duplicate column 3 in Figure 2c). Vertical lines indicate refilling of replicated groundwater and the scheduled phosphate concentration (mg/L).

In the oxic tests with phosphate amendments (columns 2 and 3) concentrations of around 500 µg/L arsenic were desorbed at the beginning, which was predominantly As(V) (Figures 1b and 2c) and declined to about 300 µg/L at the end of the test. The first addition of phosphate (4 mg/L) led to a rapid release with concentrations up to 690 µg/L and 820 µg/L arsenic respectively, more than 50%

As(III). With every phosphate amendment of increased concentration, a peak in arsenic release was observed, independently of the amended phosphate concentration. After 50 PV As(V) became the dominant species and remained the only species until the end of the test. In columns 2 and 3 the arsenic concentration of the outflow decreased to 35 µg/L by the end of the experiment. In total 4.69 mg and 4.37 mg arsenic were released from columns 2 and 3 respectively (Table 1).

Table 1. Arsenic contents in mg in aqua regia extractions of the sediment before and after the column tests in comparison with the cumulative release measured from the column outflow.

Column Test	1	2	3	4	5	6
Cumulative outflow release (mg)	0.67 *	4.69	4.37	3.20	11.4	13.1
Sedimentary As before test (mg)		17 ± 5			25 ± 0.8	
Sedimentary As after test (mg)		12.1 ± 2.8	11.4 ± 3.8		15.5 ± 3.3	12.9 ± 2.6
Sedimentary mass losses (mg)		4.9 ± 7.8	5.6 ± 8.8		9.5 ± 4.1	12.1 ± 3.4

* Shorter duration of test.

Aqua regia extractions before and after the column test show a sedimentary loss of arsenic similar to the cumulative release during the test, which corresponds to a total arsenic decrease of 26.5%. Accordingly, the arsenic release of columns 2 and 3 with phosphate amendment was six times higher than in the reference column 1 without phosphate addition. Of the added phosphate, 93 to 97% could be retrieved in total from the column outflow.

In the oxic column, the first arsenic release peak is observed together with high manganese concentrations. Nitrate was not detectable, iron and sulphate remain low (Figures 3 and 2) and manganese returns to low levels after 55 PV, while arsenic release is continuously observed in accordance with phosphate amendments (Figure 2).

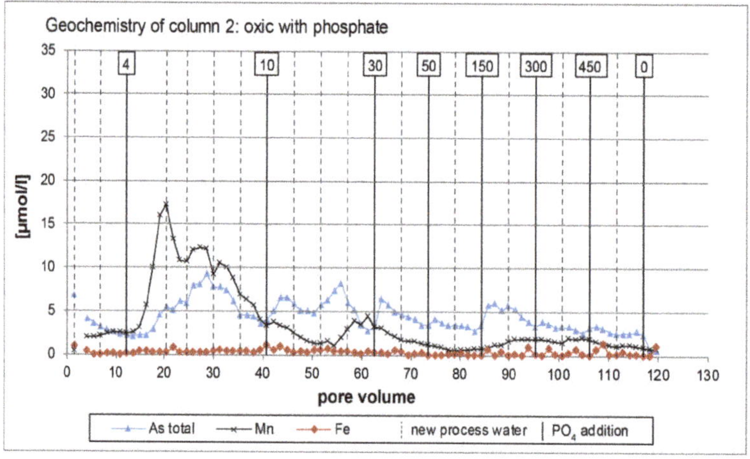

Figure 2. Release of arsenic, manganese and iron in the oxic column 2. Arsenic release is independent from geochemical milieu conditions, although it is temporarily accompanied with elevated manganese.

3.2. Anoxic Tests

In the anoxic column without phosphate addition (column 4, Figure 3a) the initial concentration of arsenic in the outflow was around 200 µg/L and increased to about 300 µg/L during the test. After 10 PV the species distribution shifted from predominantly As(V) in the beginning to almost exclusively As(III), which remained the main species until the end of the test. A total of 3.20 mg arsenic was released from the column.

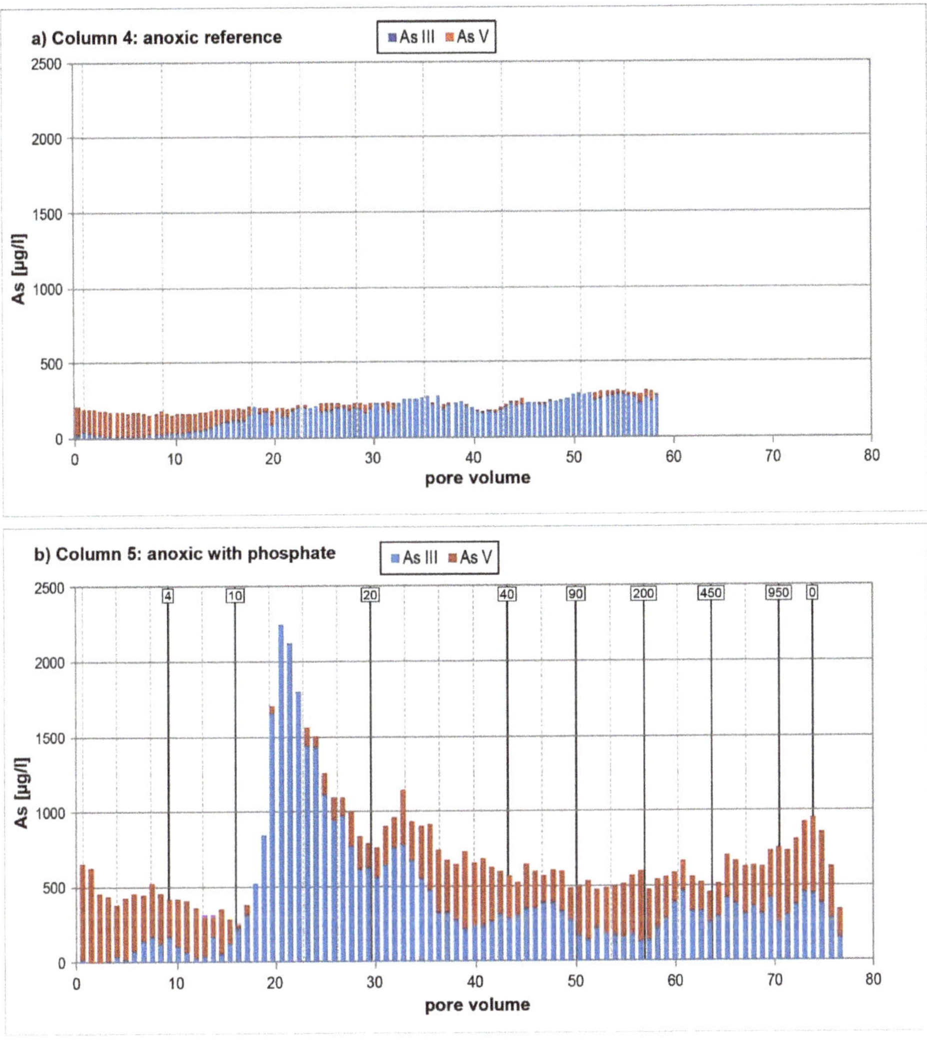

Figure 3. Arsenic release during the anoxic column tests, (**a**) column 4 without phosphate addition and (**b**) column 5 with phosphate addition. Vertical lines indicate refilling of replicated groundwater and the scheduled phosphate concentration (mg/L).

Both anoxic columns 5 and 6 (Figures 3b and 4c) at the beginning show an arsenic release of around 600 µg/L, which is predominantly As(V). About 10 PV after the addition of 10 mg/L phosphate, the arsenic release increased significantly up to 2251 µg/L and decreased again immediately. Every increase in added phosphate concentration was followed by an arsenic release peak. The amplitudes of the following peaks were lower than the first mobilization peak, even with very high phosphate dosages. According to the release balances, most of the extractable arsenic was mobilized during the first two peaks.

The species distribution in both anoxic columns with phosphate amendment shows a similar pattern. Differences are caused by heterogeneities of the sediment like variations in arsenic concentrations and organic compounds. With the onset of the first peak the species distribution changed from initially predominantly As(V) towards almost exclusively As(III). In column 5 the shares of As(V) increased again after 30 PV and from 42 PV until at the end the distribution of both species stayed equal. In contrast, As(III) remains the dominating species in column 6 after the first release peak. The high release concentrations of arsenic correlated with high shares of As(III) in both anoxic columns.

Total arsenic amounts of 11.4 mg (column 5) and 13.1 mg (column 6) were released during the anoxic tests (Table 1). According to a sediment concentration of 25 mg/kg (aqua regia extraction) before the test, 44–52% of the total arsenic content was released from the columns, which was validated by the aqua regia extractions after the column tests.

Through the phosphate amendment the arsenic release could be accelerated significantly and enhanced by four times on average in comparison to the reference. 85% of the amended phosphate was retrieved at the column outflow.

In the anoxic columns (5 and 6) the onset of iron reducing conditions were observed after 30 PV and thus after the main arsenic peak (Figures 4 and 4). Strongly iron reducing conditions with high iron concentrations in the outflowing water did not correlate with high arsenic concentrations. Sulphate concentrations showed no change with time but from the added sulphate in the process water up to 7.72% was consumed in the column (Table 7). The constant HCO_3 concentrations in the column outflow indicate stable pH values as detected in preliminary studies (Figure 5). Manganese increased slightly throughout the test after phosphate amendments.

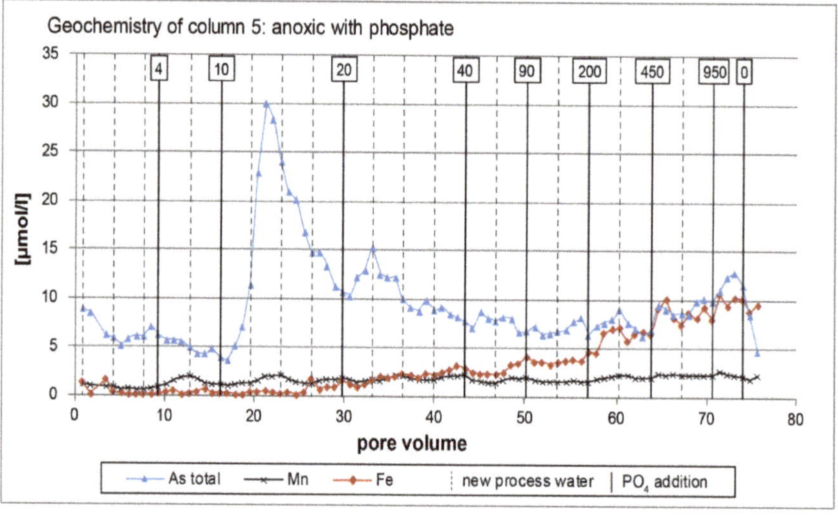

Figure 4. Release of arsenic, manganese and iron in the anoxic column 5. Manganese increases steadily while iron remains low and increases right after the high arsenic release peak.

4. Discussion

In all columns with phosphate amendment a significant release of arsenic was observed from the beginning during low phosphate amendment concentrations. Even smaller amounts of phosphate than used in other studies (0.04 mM instead of 6 mM) led to a comparable amount of arsenic release [35]. Phosphate concentrations measured from the column outflow demonstrate that most of the added phosphate was neither adsorbed or precipitated in the columns nor consumed by microorganisms.

As expected the total arsenic release rates were higher and faster under anoxic conditions. However, in comparison to the reference the release rates in columns 2 and 3 with phosphate were significantly increased under oxic conditions. Consequently, the addition of phosphate increased microbial processes, which consumed oxygen and nitrate, temporarily causing manganese-reducing conditions. This observed manganese release occurred simultaneously with a small peak of arsenic. Groundwater composition does not indicate a further development of suboxic conditions after the manganese peak but rather a return to more oxic conditions with low iron, low manganese and predominantly As(V). In the anoxic columns with phosphate (columns 5 and 6), the released arsenic contains high shares of both arsenic species compared to the reference (column 4). This shows that phosphate amendment significantly increased the normal arsenic desorption process under anoxic conditions. High mobility of As(V) can be attributed to competitive surface complexation with phosphate under both applied conditions, while the pattern of arsenic species might be attributed to microbial processes. The increased arsenic release is independent of the prevailing redox conditions and no significant effect on the pH value was observed (Figure 5).

Accelerated microbial activity in the columns with phosphate addition is indicated by the loss of sulphate (Table 7) and the increased iron and manganese in the column outflow. In both reference columns without phosphate, the sulphate concentration remained stable and no significant amounts of iron or manganese were released (Figure 6).

At a neutral pH value, the release of other metals besides arsenic is not probable. Before the field application of this approach at other sites without high carbonate buffering the possibility of heavy metal release should be tested. The aquifer material of the study area contained no other metals besides arsenic, thus the effect of phosphate amendments on other metals could not be considered.

Variations between the cumulative outflow and the sedimentary mass balances are caused by the different methodological approaches and the standard deviation of sedimentary arsenic measurements (Table 2). In the column outflows with phosphate amendment the arsenic release is 3.5 to 4.5 times higher under oxic and anoxic conditions. Compared to the phosphate extractable fractions in Table 2, a total of 36% (oxic) to 65% (anoxic) arsenic was released, most of it already with low additions of phosphate. The significant shift of arsenic species under anoxic conditions from As(V) to As(III) in the column outflow indicates enhanced microbial arsenic reduction in the column, as in the phosphate elution most of the extractable and thus surface bound arsenic was As(V).

Table 2. Arsenic species distribution in the sediment elutions and the release measured from the column outflow.

Column	Water Extractable Arsenic (mg)		Phosphate (1 M) Extractable Arsenic (mg)		Arsenic Released from Column (mg)	
	As(III)	As(V)	As(III)	As(V)	As(III)	As(V)
2 (oxic)	0.13	1.24	1.19	11.9	1.13	3.58
5 (anoxic)	0.20	2.24	1.48	16.3	6.78	4.60

The first phosphate amendments with low concentrations (10 mg/L) triggered a strong As(III) release. After this first peak, As(V) shares increased with respect to redox conditions. These findings indicate that a complex interplay of microbial processes and surface complexation govern the distribution of arsenic species depending on the availability of phosphate (Figure 5).

Preliminary tests indicated that under all described conditions a supersaturated phosphate concentration is necessary to start the release process [36]. The column tests in this study show that rapid release already starts at relatively low phosphate concentrations of 4 mg/L (0.04 mM) and continues when concentrations are increased during time. The wide range of phosphate amendment concentrations used in the other published studies illustrate that processes not only rely on the concentration of phosphate but are also strongly linked to the geochemistry of the sediment matrix, the water composition and microbial activity.

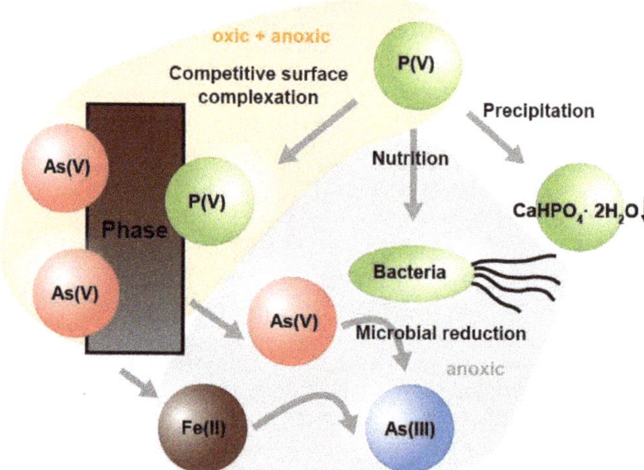

Figure 5. Schematic sketch of dominating processes supported by the results of this study: arsenic release by competitive surface complexation on iron and calcite minerals (phase), reduction of As(V) to As(III) by Fe(II) or arsenic-reducing bacteria and phosphate precipitation mainly as brushit ($CaHPO_4 \cdot 2H_2O$).

During the oxic tests, iron and manganese were not observed in significant concentrations in the column output, whereas nitrate reduction could be detected, which indicates that microbial activity in the columns leads to suboxic conditions. The temporal shift towards As(III) in the oxic tests with phosphate amendment might be artefacts of anoxic conditions in the sediment used for the columns or the fraction of As(III) on the sediment matrix (Table 2) but are more likely the effects of microbial processes fuelled by phosphate addition. Also, the high shares of As(III) in the anoxic tests were not linked to significant increases of manganese or iron. This is quite surprising considering the current standard of knowledge that microbial arsenic reduction prevails under iron-reducing conditions.

The observed rapid reduction of As(V) to As(III) after the 10 mg/L phosphate amendment could be explained by the stimulation of arsenic-reducing microorganisms by phosphate, which is likely the limiting factor for biological activity in the observed case. Macur et al. (2004) support the hypothesis that microorganisms are capable to oxidize and to reduce arsenic coexist in natural sediments [44]. Both tests indicate that the microorganisms were present in the sediment and just needed a time of incubation to grow in population. Campbell et al. (2006) subdivided two pathways of microbial As(V) reduction, the respiratory pathway and the detoxification [45]. Under anoxic conditions, the respiratory pathway was the dominating process. According to scientific consensus the As(V) reduction to As(III) by microbial respiration processes comes along with reducing conditions. For example, Huang et al. (2011) and Dhar et al. (2011) identified a microbe that mediated As(V) reduction under iron reducing conditions [46,47]. They agree with Zhang et al. (2009) that microbial desorption and iron reduction occur independently [48].

During our oxic and anoxic tests the As(V) release and subsequent reduction to As(III) was found independently from iron reducing conditions, which could be explained with the competitive surface complexation by phosphate from iron or carbonate surfaces. Another reason could be the dissolution of arsenic-bearing carbonates [28,30]. We suggest that the arsenic reduction of As(V) to As(III) is mainly due to microbial detoxification and is thus decoupled from the redox condition. The increased availability of phosphate leads to accelerated microbial activity combined with the uptake of As(V) in competition with phosphate and the release of As(III) species after the detoxification process.

In addition, Slaugther et al. (2012) demonstrated that arsenate reduction is suppressed at phosphate concentrations > 0.5 mM and even inhibited at high P/As ratios [49]. This observation gives a reasonable explanation for the decrease of these microbial detoxification processes with increasing phosphate amendments and a shift of arsenic species from dominantly As(III) to As(V) during the experiments. Further research will be necessary to understand the ongoing microbial processes.

5. Conclusions

The presented column tests clearly show that even low concentrations of phosphate amendment cause competitive exchange and can enhance biological processes, while higher phosphate dosages suppress microbial activity. Results also indicate that during low amendments of phosphate the microbial reduction of As(V) occurs very fast and is decoupled from manganese or iron reduction.

Enhanced in-situ arsenic remediation with phosphate bears great potential to accelerate the remediation process considerably. It is suitable for oxic and anoxic redox conditions, as it has no strong effects onto the prevailing redox milieu. Side effects such as mobility of heavy metals were not investigated in this study but have to be considered when applying the method at other places.

Additionally, transport velocities in field applications might be slower due to lower effective porosity of diagenetically compressed sediments in comparison with the manually compacted sediments in the columns. Furthermore, like in all in-situ remediation approaches, the heterogeneities in the sediment profile can change the effect and efficiency of the remediation, which could not be evaluated from the homogenized column material in this study.

Regarding the high phosphate amendment in total, the shares of phosphate consumed during the release processes are very low. Before field application, reactions with excess phosphate have to be considered. In the carbonate-rich aquatic environment of this study, the phosphate is precipitated with calcium, forming mainly white crystalline brushit $CaHPO_4 \cdot 2H_2O$ (Figures 5 and 5) and shares of hydroxyl apatite $Ca_5(PO_4)_3(OH)$.

In carbonate-dominated systems, co-reactions of phosphate need to be compensated by a higher dosage. An economical application of phosphate is aspired for economic and ecologic reasons. Therefore, an optimal injection technique should be used in order to minimize phosphate amendments for efficient implementation in the remediation practice. Only in the anoxic columns significant amounts of phosphate remained in the sediment matrix after test termination. Future investigations should consider the phosphate precipitation and a potential field application [50].

In order to understand all occurring processes in the aquifer, larger scale studies in the field and geochemical modelling on the phosphate-induced arsenic release should be obtained in the framework of future project cooperation.

Author Contributions: Conceptualization, methodology, formal analysis, investigation, project administration, M.V.M.; methodology, validation and formal analysis, C.S.; supervision: M.I.-S.; writing—original draft preparation and visualization, Y.W. and D.Z.; writing—review, C.N.S.; writing—review and editing, M.V.M. and M.I.-S.

Funding: This work was funded by HIM ASG, Biebesheim and the regional council of Darmstadt (RP Darmstadt). We acknowledge financial support by Deutsche Forschungsgemeinschaft within the funding programme Open Access Publishing, by the Baden-Württemberg Ministry of Science, Research and the Arts and by Ruprecht-Karls-Universität Heidelberg.

Acknowledgments: The authors like to thank all colleagues from the University of Heidelberg, who contributed to the studies with laboratory works and scientific input. We also thank CDM Smith for technical and the Hessian Agency for Nature Conservation, Environment and Geology (HNLUG) for consultant support. Thank you to Amanda Wendt and Susanne Braun for language corrections and the reviewers for the helpful remarks and feedback.

Conflicts of Interest: The authors declare no conflict of interest. The funders had no role in the design of the study; in the collection, analyses or interpretation of data; in the writing of the manuscript or in the decision to publish the results.

Appendix A

Table A1. Groundwater composition at the study site.

			Field Parameters				
Sample ID	Sampling Depth	Temperature	Electrical Conductivity	pH Value	Acid Capacity	ORP	Dissolved Oxygen
	(m bgs)	(°C)	(µS/cm)	(-)	(mmol)	(mV)	
GWM 29/1	7.80	13.3	795	7.34	4.6	149	1.1
GWM 29/2	15.50	12.5	617	7.42	4.1	49	0.2
GWM 29/3	22.50	13.1	646	7.39	4.1	117	0.8

				Major cations					
Sample ID	Sampling Depth	Fe	Mn	As	As III	As V	Ca	K	Mg	Na
	(m bgs)	(mg/L)	(µg/L)	(µg/L)	(µg/L)	(µg/L)	(mg/L)	(mg/L)	(mg/L)	(mg/L)
GWM 29/1	7.80	2.24	407	667	19.0	648	154	9.46	9.74	18.9
GWM 29/2	15.50	1.41	462	322	68.1	254	120	6.29	8.51	12.1
GWM 29/3	22.50	1.21	480	230	26.9	203	129	7.21	9.02	14.4

				Heavy metals					
Sample ID	Sampling Depth	Al	Cd	Co	Cr	Cu	Ni	Pb	Sr	Zn
	(m bgs)	(µg/L)	(µg/L)	(µg/L)	(µg/L)	(µg/L)	(µg/L)	(µg/L)	(µg/L)	(µg/L)
GWM 29/1	7.80	<50	<5	<5	<5	<5	<5	<10	244	60
GWM 29/2	15.50	<50	<5	<5	<5	<5	<5	<10	258	29.6
GWM 29/3	22.50	<50	<5	<5	<5	<5	<5	<10	260	33.5

			Anions				
Sample ID	Sampling Depth	NO$_3$	SO$_4$	PO$_4$	F	Cl	Br
	(m bgs)	(mg/L)	(mg/L)	(mg/L)	(mg/L)	(mg/L)	(mg/L)
GWM 29/1	7.80	5.4	171	<1	0.50	20.5	0.08
GWM 29/2	15.50	1.1	106	<1	0.20	14.9	<0.01
GWM 29/3	22.50	1.2	126	<1	0.27	16.4	0.04

			Carbon		
Sample ID	Sampling Depth	HCO$_3{}^-$	TC	TIC	TOC
	(m bgs)	(mg/L)	(mg/L)	(mg/L)	(mg/L)
GWM 29/1	7.80	270	58.9	53	5.87
GWM 29/2	15.50	225	49.7	44.2	5.51
GWM 29/3	22.50	247	52.3	48.5	3.75
GWM 29/3	22.50	247	52.3	48.5	3.75

Table 2. Overview of the used measurements methods, ranges and quality control.

Method	Limit of Determination (As)	Reference Material	Standard Deviation
ICP-OES (water)	50 μg/L	SPW-SW2 TMDA 51.3	1–15%
HG-ICP-OES (As(III))	2.50 μg/L	3× replication of measurement	1–8%
ICP-OES (aqua regia/elutions)	2.50 mg/kg	BAM U112a, Blank	2.83–5.35%
Photometer (phosphate)	0.53 mg/L		6.02%

Table 3. Geochemical characteristics of the aquifer material from the liner (GWM 29) used for the columns, results from aqua regia extractions (n.n. = below detection).

Sample	Depth	As Total	Ca	Mn	Fe	CaCO$_3$	Carbonate	C Total	C$_{org}$ (calc.)
	(m bgs)	(mg/kg)	(%)	(mg/kg)	(%)	(%)	(as %C)	(%)	(%)
29-1 + 2	1–2	35.2	0.56	151	0.63	n.n.	n.n.	n.n.	n.n.
29-3	3	12.5	1.82	105	0.45	4.30	0.52	0.64	0.12
29-5	5	60.5	0.72	28	0.20	2.15	0.28	0.23	0
29-7	7	25.8	1.86	56.2	0.27	6.46	0.78	0.96	0.18
29-9	9	28.9	3.30	68.2	0.29	7.94	0.95	1.15	0.20
29-11	11	33.6	2.95	85	0.41	12.2	1.47	1.36	0
29-13	13	22.2	2.17	59.7	0.30	7.27	0.87	0.70	0
29-15	15	25.4	2.87	66.2	0.31	6.13	0.74	0.77	0.03

Table 4. Composition of the replicated groundwater used for the column tests based on the water composition at the contaminated site.

Ion	Concentration (mg/L)	Added as	Source and Quality
Ca^{2+}	128	CaSO$_4$$^{2-}$·H$_2$O and CaCO$_3$	MERCK/Grüssing (p.a)
Mg^{2+}	8	MgCl$_2$$^{2-}$·6H$_2$O	AppliChem (p.a)
Na$^+$	10	NaHCO$_3$	Grüssing (p.a)
K$^+$	5	KHCO$_3$	Theoretikum (p.a)
NO$_3$$^-$	5 *	NaNO$_3$	Grüssing (p.a)
SO$_4$$^{2-}$	120	CaSO$_4$$^{2+}$	MERCK (p.a)
Cl$^-$	23	MgCl$_2$$^{2-}$·6H$_2$O	AppliChem (p.a)

* Only for oxic columns, for anoxic columns: 0 mg/L.

Table 5. Column geometry, redox conditions and phosphate application.

Column	Diameter (m)	Length (m)	Redox Condition	Discharge (L/day)	Phosphate Application
1	0.005	0.3	Oxic	0.18	none
2	0.004	0.3	Oxic	0.22	via bypass into
3	0.004	0.3	Oxic	0.22	columns entry
4	0.005	0.3	Anoxic	0.18	none
5	0.005	0.3	Anoxic	0.18	added to process water
6	0.005	0.3	Anoxic	0.18	after degassing

Table 6. Considered parameters, applied sample preparation and analytical methods.

Sediment Samples

Aqua regia extraction	As (tot) and cations	1 g sample with 3 mL 65%, HNO_3 and 9 mL 37% HCl, filled up to 50 mL and filtered through folded filter	(HG)-ICP-OES (VISTA-MPX CCD Simultaneous; Varian)
Water elution	As (III) and As (tot)	3.5 g sample add. 40 mL distilled water, filtered by 0.45 μm and stabilized at pH 1 with 6 M HCl	
Phosphate elution	As (III) and As (tot)	3.5 g sample add. 40 mL 1 M NaH_2PO_4, filtered by 0.45 μm and stabilized at pH 1 with 6 M HCl	
Solid sample	TC		SC-Analyzer (SC 144R; Leco)

Water Samples

Inflow solution and column outflow	As (III), As (tot) and cations	stabilized at pH 1 with 6 M HCl	(HG)-ICP-OES (VISTA-MPX CCD Simultaneous; Varian)
	Anions		IC (Dionex DX-120; ThermoFisher Scientific)
	TIC/TC		TOC (TOC-VCSN; Shimadzu)
	Phosphate	Molybdenum-blue method	UV-VIS Spectrophotometer; (Specord 50; Analytik Jena)
Tracer test	Bromide		Ion selective electrode (Br 500; WTW); IC (Dionex DX-120; ThermoFisher Scientific)

Table 7. Sulphate balancing from the inflowing (replicated water) and outflowing water.

		Average SO_4 (mg/L)			
		Replicated Water		Column Outflow	Deviation SO_4 (%)
	Column	Planned	Measured *		
1	reference oxic		125	129	3.25
2	oxic + PO_4	120	120	111	−7.72
3	oxic + PO_4			115	−4.71
4	reference anoxic		120	121	0.17
5	anoxic + PO_4		121	113	−6.89
6	anoxic + PO_4			114	−6.51

* Deviations from the planned amount are due to statistical outliers.

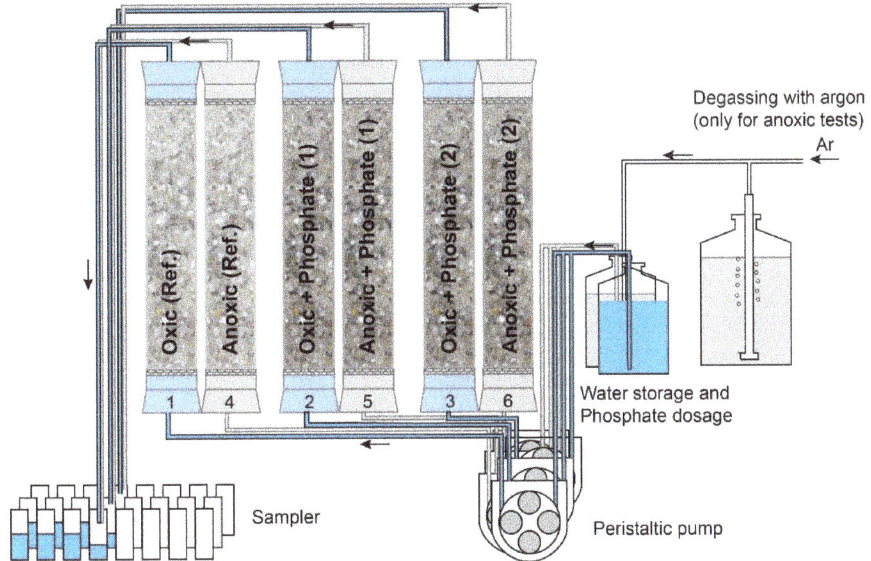

Figure 1. Setup of the column tests. Columns filled with contaminated sediment are percolated from bottom to top by replicated groundwater, to which phosphate is dosed stepwise. For the anoxic tests, the water was degassed with argon before percolation and the whole system was run in an argon floated glove box (modified after [51]).

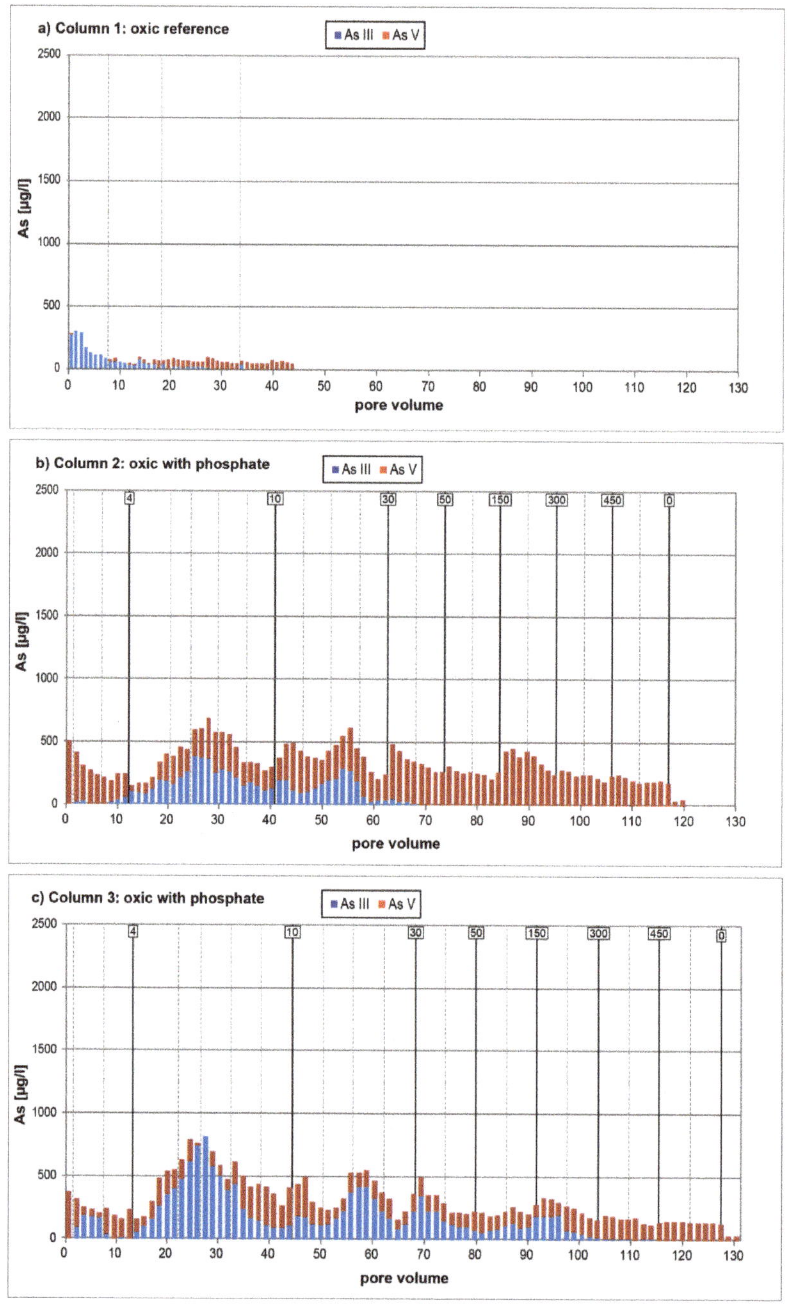

Figure 2. Arsenic release during the oxic column tests, (**a**) column 1 without phosphate addition and (**b**) + (**c**) with phosphate addition in column 2 and 3. Vertical lines indicate refilling of replicated groundwater and the numbers next to them the scheduled phosphate concentration (mg/L).

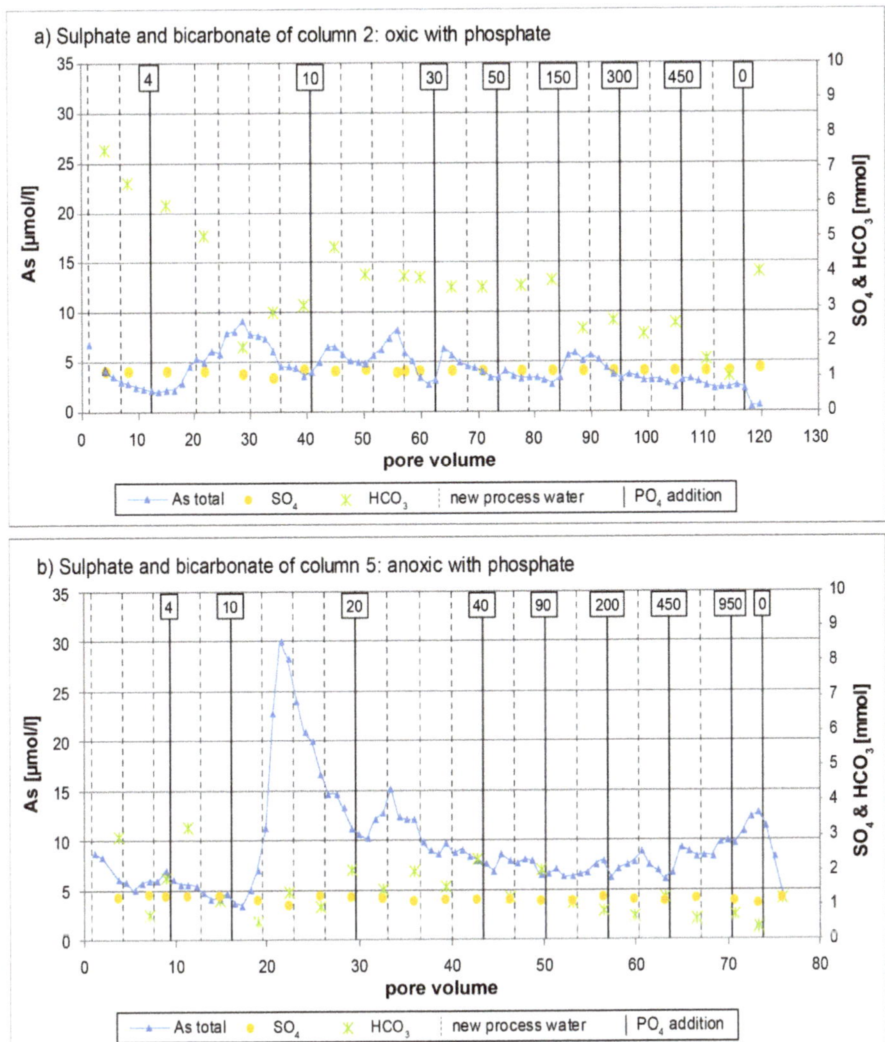

Figure 3. Sulphate (SO_4) and bicarbonate (HCO_3) concentrations (right axis) in comparison with arsenic (left axis) in (**a**) oxic column 2 and (**b**) anoxic column 5.

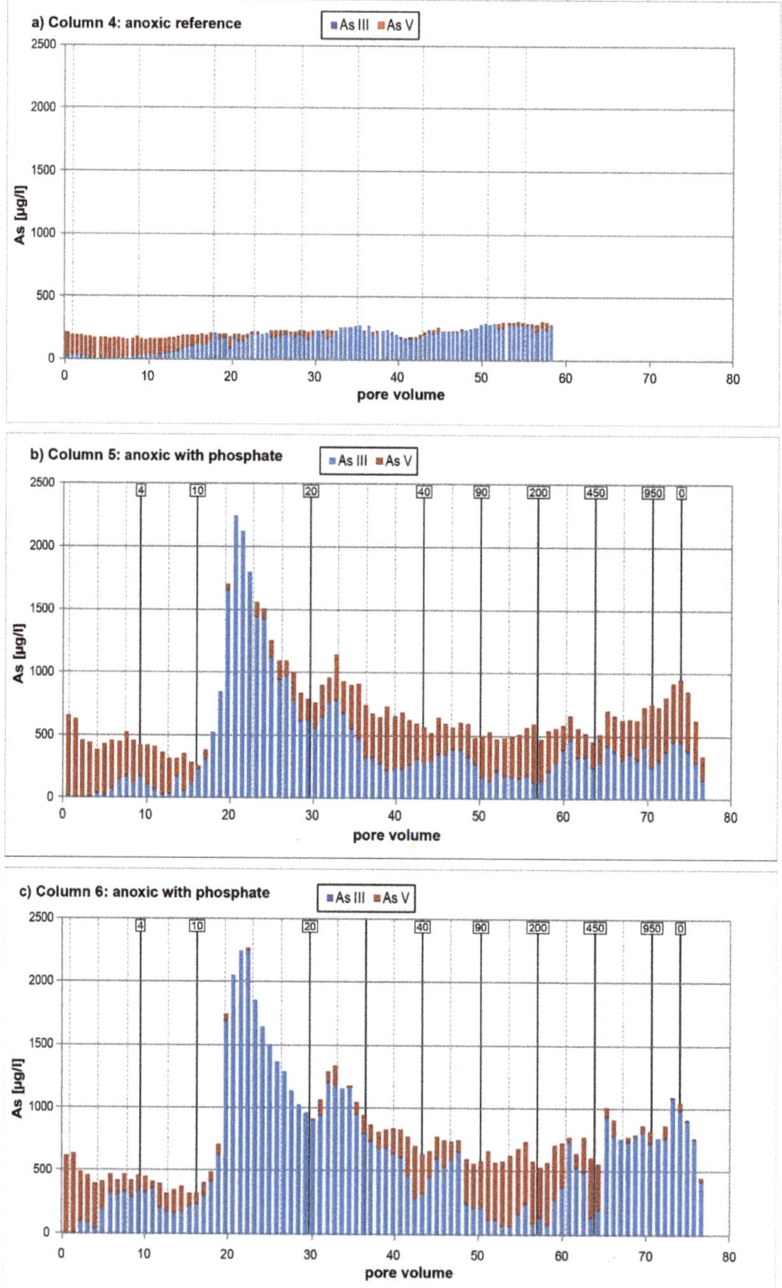

Figure 4. Arsenic release during the anoxic column tests, (**a**) column 4 without phosphate addition and (**b**) + (**c**) column 5 and 6 with phosphate addition. Vertical lines indicate refilling of replicated groundwater, the attached numbers the amended phosphate concentrations (mg/L) at the time of solution preparation.

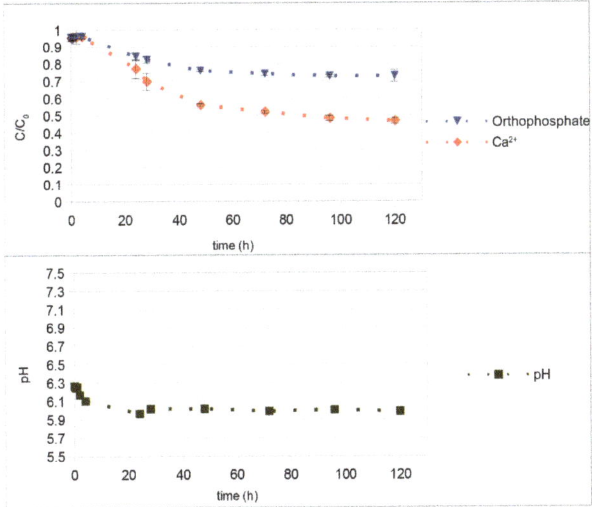

Figure 5. Bottom: Results from a batch experiment with groundwater from the study site verify that even very high phosphate amendments of 10 mmol/L only result in a small change in pH value from pH 6.3 to 6 without any further decline. Top: Change in concentration (c/c_0) of calcium and orthophosphate over 120 h show the precipitation of brushit [52].

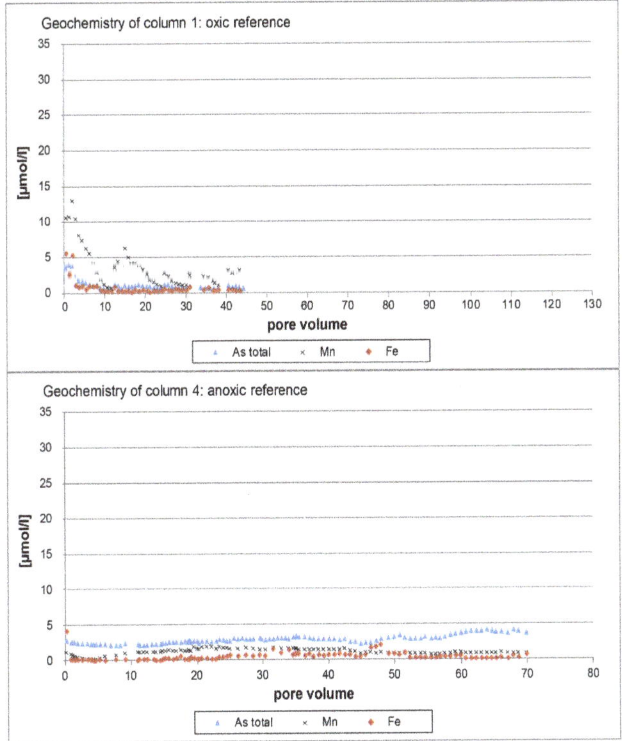

Figure 6. Release of arsenic, manganese and iron in the oxic reference column 1 (top) and the anoxic reference column 4 (bottom).

References

1. Eljamal, O.; Sasaki, K.; Tsuruyama, S.; Hirajima, T. Kinetic Model of Arsenic Sorption onto Zero-Valent Iron (ZVI). *Water Qual. Expo. Health* **2011**, *2*, 125–132. [CrossRef]
2. Bissen, M.; Frimmel, F.H. Arsenic—A Review. Part II: Oxidation of Arsenic and its Removal in Water Treatment. *Acta Hydrochim. Hydrobiol.* **2003**, *31*, 97–107. [CrossRef]
3. Kuhlmeier, P.D. Sorption and desorption of arsenic from sandy soils: Column studies. *J. Soil Contam.* **1997**, *6*, 21–36. [CrossRef]
4. Smedley, P.L.; Kinniburgh, D.G. A review of the source, behaviour and distribution of arsenic in natural waters. *Appl. Geochem.* **2002**, *17*, 517–568. [CrossRef]
5. Guo, H.; Stüben, D.; Berner, Z.; Yu, Q. Characteristics of arsenic adsorption from aqueous solution: Effect of arsenic species and natural adsorbents. *Appl. Geochem.* **2009**, *24*, 657–663. [CrossRef]
6. Bowell, R.J. Sorption of arsenic by iron oxides and oxyhydroxides in soils. *Appl. Geochem.* **1994**, *9*, 279–286. [CrossRef]
7. Dzombak, D.A.; Morel, F.M.M. *Surface Complexation Modeling: Hydrous Ferric Oxide*; Wiley-Interscience: New York, NY, USA, 1990.
8. Sø, H.U.; Postma, D.; Jakobsen, R.; Larsen, F. Sorption of phosphate onto calcite; results from batch experiments and surface complexation modeling. *Geochim. Cosmochim. Acta* **2011**, *75*, 2911–2923. [CrossRef]
9. Biswas, A.; Majumder, S.; Neidhardt, H.; Halder, D.; Bhowmick, S.; Mukherjee-Goswami, A.; Kundu, A.; Saha, D.; Berner, Z.; Chatterjee, D. Groundwater chemistry and redox processes: Depth dependent arsenic release mechanism. *Appl. Geochem.* **2011**, *26*, 516–525. [CrossRef]
10. Saunders, J.A.; Lee, M.-K.; Shamsudduha, M.; Dhakal, P.; Uddin, A.; Chowdury, M.T.; Ahmed, K.M. Geochemistry and mineralogy of arsenic in (natural) anaerobic groundwaters. *Appl. Geochem.* **2008**, *23*, 3205–3214. [CrossRef]
11. Rochette, E.A.; Bostick, B.C.; Li, G.; Fendorf, S. Kinetics of Arsenate Reduction by Dissolved Sulfide. *Environ. Sci. Technol.* **2000**, *34*, 4714–4720. [CrossRef]
12. Bostick, B.C.; Fendorf, S. Arsenite sorption on troilite (FeS) and pyrite (FeS$_2$). *Geochim. Cosmochim. Acta* **2003**, *67*, 909–921. [CrossRef]
13. Bundschuh, J.; Hollander, H.; Lena, M. *In-Situ Remediation of Arsenic-Contaminated Sites*, 1st ed.; Bundschuh, J., Hollander, H., Lena, M., Eds.; CRC Press, Taylor & Francis Group: London, UK, 2014.
14. Matthess, G. In Situ Treatment of Arsenic Contaminated Groundwater. In *Studies in Environmental Science*; Elsevier: Amsterdam, The Netherlands, 1981; Volume 17, pp. 291–296. [CrossRef]
15. Rott, U.; Friedle, M. Eco-friendly and cost-efficient removal of arsenic, iron and manganese by means of subterranean ground-water treatment. *Water Supply* **2000**, *18*, 632–636.
16. Kim, M.-J.; Nriagu, J. Oxidation of arsenite in groundwater using ozone and oxygen. *Sci. Total Environ.* **2000**, *247*, 71–79. [CrossRef]
17. Krüger, T.; Holländer, H.M.; Stummeyer, J.; Harazim, B.; Boochs, P.-W.; Billib, M. In-situ immobilization of arsenic in the subsurface on an anthropogenic contaminated site. In *In-Situ Remediation of Arsenic-Contaminated Sites*; Bundschuh, J., Holländer, H.M., Ma, L.Q., Eds.; CRC Press, Taylor & Francis Group: London, UK, 2014.
18. Luong, V.T.; Cañas Kurz, E.E.; Hellriegel, U.; Luu, T.L.; Hoinkis, J.; Bundschuh, J. Iron-based subsurface arsenic removal technologies by aeration: A review of the current state and future prospects. *Water Res.* **2018**, *133*, 110–122. [CrossRef] [PubMed]
19. Onstott, T.C.; Chan, E.; Polizzotto, M.L.; Lanzon, J.; DeFlaun, M.F. Precipitation of arsenic under sulfate reducing conditions and subsequent leaching under aerobic conditions. *Appl. Geochem.* **2011**, *26*, 269–285. [CrossRef]
20. Gemeinhardt, C.; Müller, S.; Weigand, H.; Marb, C. Chemical immobilisation of arsenic in contaminated soils using iron(II)sulphate—advantages and pitfalls. *Water Air Soil Pollut. Focus* **2006**, *6*, 281–297. [CrossRef]
21. Köber, R.; Daus, B.; Ebert, M.; Mattusch, J.; Welter, E.; Dahmke, A. Compost-Based Permeable Reactive Barriers for the Source Treatment of Arsenic Contaminations in Aquifers: Column Studies and Solid-Phase Investigations. *Environ. Sci. Technol.* **2005**, *39*, 7650–7655. [CrossRef] [PubMed]
22. Klaas, N.; Braun, J.; Mackenberg, S. *Wissenschaftlicher Bericht Nr. 2007/06 (VEG 23) Entwicklung Eines Immobilisierungsverfahrens für Schwermetalle unter Nutzung des Geogenen Sulfatgehaltes im Grundwasser*; VEGAS: Stuttgart, Germany, 2007.

23. Zeng, H.; Fisher, B.; Giammar, D.E. Individual and Competitive Adsorption of Arsenate and Phosphate to a High-Surface-Area Iron Oxide-Based Sorbent. *Environ. Sci. Technol.* **2008**, *42*, 147–152. [CrossRef] [PubMed]

24. Kent, D.B. The influence of groundwater chemistry on arsenic concentrations and speciation in a quartz sand and gravel aquifer. *Geochem. Trans.* **2004**, *5*, 1–12. [CrossRef]

25. Hongshao, Z.; Stanforth, R. Competitive Adsorption of Phosphate and Arsenate on Goethite. *Environ. Sci. Technol.* **2001**, *35*, 4753–4757. [CrossRef] [PubMed]

26. Luengo, C.; Brigante, M.; Avena, M. Adsorption kinetics of phosphate and arsenate on goethite. A comparative study. *J. Colloid Interface Sci.* **2007**, *311*, 354–360. [CrossRef] [PubMed]

27. Manning, B.A.; Goldberg, S. Modeling Competitive Adsorption of Arsenate with Phosphate and Molybdate on Oxide Minerals. *Soil Sci. Soc. Am. J.* **1996**, *60*, 121–131. [CrossRef]

28. Bardelli, F.; Benvenuti, M.; Costagliola, P.; Di Benedetto, F.; Lattanzi, P.; Meneghini, C.; Romanelli, M.; Valenzano, L. Arsenic uptake by natural calcite: An XAS study. *Geochim. Cosmochim. Acta* **2011**, *75*, 3011–3023. [CrossRef]

29. Yokoyama, Y.; Tanaka, K.; Takahashi, Y. Differences in the immobilization of arsenite and arsenate by calcite. *Geochim. Cosmochim. Acta* **2012**, *91*, 202–219. [CrossRef]

30. Sø, H.U.; Postma, D.; Jakobsen, R.; Larsen, F. Sorption and desorption of arsenate and arsenite on calcite. *Geochim. Cosmochim. Acta* **2008**, *72*, 5871–5884. [CrossRef]

31. Pigna, M.; Krishnamurti, G.S.R.; Violante, A. Kinetics of Arsenate Sorption–Desorption from Metal Oxides. *Soil Sci. Soc. Am. J.* **2006**, *70*, 2017–2027. [CrossRef]

32. Alam, M.G.; Tokunaga, S.; Maekawa, T. Extraction of arsenic in a synthetic arsenic-contaminated soil using phosphate. *Chemosphere* **2001**, *43*, 1035–1041. [CrossRef]

33. Woolson, E.A.; Axley, J.H.; Kearney, P.C. The Chemistry and Phytotoxicity of Arsenic in Soils: II. Effects of Time and Phosphorus. *Soil Sci. Soc. Am. J.* **1973**, *37*, 254–259. [CrossRef]

34. Wovkulich, K.; Mailloux, B.J.; Lacko, A.; Keimowitz, A.R.; Stute, M.; Simpson, H.J.; Chillrud, S.N. Chemical treatments for mobilizing arsenic from contaminated aquifer solids to accelerate remediation. *Appl. Geochem.* **2010**, *25*, 1500–1509. [CrossRef] [PubMed]

35. O'Reilly, S.E.; Strawn, D.G.; Sparks, D.L. Residence Time Effects on Arsenate Adsorption/Desorption Mechanisms on Goethite. *Soil Sci. Soc. Am. J.* **2001**, *65*, 67. [CrossRef]

36. Maier, M. *Untersuchungen zum Reaktiven Transport von Arsen im Grundwasserleiter: Prozessstudie und Entwicklung Einer Neuartigen Sanierungsmethode an Einem Altstandort in Hessen*; Ruprecht-Karls-Universität Heidelberg: Heidelberg, Germany, 2014. [CrossRef]

37. Maier, M.V. Insitu-Mobilisierung von Arsen im Grundwasser. In *Handbuch Altlastensanierung und Flächenmanagement (HdA)*; Franziskus, V., Altenbockum, M., Gerhold, T., Eds.; Hütig Jehle Rehm Verlag: Heidelberg, Germany, 2016; ISBN 978-3-8073-2397-8.

38. Keon, N.E.; Swartz, C.H.; Brabander, D.J.; Harvey, C.; Hemond, H.F. Validation of an arsenic sequential extraction method for evaluating mobility in sediments. *Environ. Sci. Technol.* **2001**, *35*, 2778–2784. [CrossRef] [PubMed]

39. Rüde, T.R. *Beiträge zur Geochemie des Arsens*; Puchelt, H., Ed.; Karlsruher Geochemische Hefte: Karlsruhe, Germany, 1996; Volume 10.

40. Höhn, R.; Isenbeck-Schröter, M.; Kent, D.B.; Davis, J.A.; Jakobsen, R.; Jann, S.; Niedan, V.; Scholz, C.; Stadler, S.; Tretner, A. Tracer test with As(V) under variable redox conditions controlling arsenic transport in the presence of elevated ferrous iron concentrations. *J. Contam. Hydrol.* **2006**, *88*, 36–54. [CrossRef] [PubMed]

41. Appelo, C.A.J.; Postma, D. *Geochemistry, Groundwater and Pollution*, 2nd ed.; CRC Press, Taylor & Francis Group: Boca Raton, FL, USA; London, UK; New York, NY, USA, 2005.

42. Kinzelbach, W. *Numerische Methoden zur Modellierung des Transports von Schadstoffen im Grundwasser*; 2. Aufl.; Oldenbourg: München, Germany; Wien, Austria, 1987.

43. Isenbeck-Schröter, M. *Transportverhalten von Schwermetallkationen und Oxoanionen–Laborversuche in Säulen und ihre Modellierung*; Berichte Nr. 67; Fachbereich Geowissenschaften: Bremen, Germany, 1995.

44. Macur, R.E.; Jackson, C.R.; Botero, L.M.; Mcdermott, T.R.; Inskeep, W.P. Bacterial Populations Associated with the Oxidation and Reduction of Arsenic in an Unsaturated Soil. *Environ. Sci. Technol.* **2004**, *38*, 104–111. [CrossRef] [PubMed]

45. Campbell, K.M.; Malasarn, D.; Saltikov, C.W.; Newman, D.K.; Hering, J.G. Simultaneous Microbial Reduction of Iron(III) and Arsenic(V) in Suspensions of Hydrous Ferric Oxide. *Environ. Sci. Technol.* **2006**, *40*, 5950–5955. [CrossRef] [PubMed]

46. Huang, J.-H.; Voegelin, A.; Pombo, S.A.; Lazzaro, A.; Zeyer, J.; Kretzschmar, R. Influence of Arsenate Adsorption to Ferrihydrite, Goethite, and Boehmite on the Kinetics of Arsenate Reduction by Shewanella putrefaciens strain CN-32. *Environ. Sci. Technol.* **2011**, *45*, 7701–7709. [CrossRef] [PubMed]

47. Dhar, R.K.; Zheng, Y.; Saltikov, C.W.; Radloff, K.A.; Mailloux, B.J.; Ahmed, K.M.; van Geen, A. Microbes Enhance Mobility of Arsenic in Pleistocene Aquifer Sand from Bangladesh. *Environ. Sci. Technol.* **2011**, *45*, 2648–2654. [CrossRef] [PubMed]

48. Zhang, X.; Jia, Y.; Wang, S.; Pan, R.; Zhang, X. Bacterial reduction and release of adsorbed arsenate on Fe(III)-, Al- and coprecipitated Fe(III)/Al-hydroxides. *J. Environ. Sci.* **2012**, *24*, 440–448. [CrossRef]

49. Slaughter, D.C.; Macur, R.E.; Inskeep, W.P. Inhibition of microbial arsenate reduction by phosphate. *Microbiol. Res.* **2012**, *167*, 151–156. [CrossRef] [PubMed]

50. Maier, M.V.; Isenbeck-Schröter, M.; Klose, L.B.; Ritter, S.M.; Scholz, C. In situ-mobilization of arsenic in groundwater–an innovative remediation approach? *Procedia Earth Planet. Sci.* **2017**, *17*, 452–455. [CrossRef]

51. Tretner, A. *Sorptions- und Redoxprozesse von Arsen an Oxidischen Oberflächen–Experimentelle Untersuchungen*; Ruprecht-Karls-Universität Heidelberg: Heidelberg, Germany, 2002. [CrossRef]

52. Burger, J. *Untersuchungen zur Fällungskinetik von Calciumphosphaten Batchversuche mit Grundwasser*; Ruprecht-Karls-University Heidelberg: Heidelberg, Germany, 2014.

 *water* 

Article

Significance of Chlorinated Phenols Adsorption on Plastics and Bioplastics during Water Treatment

Aleksandra Tubić, Maja Lončarski *, Snežana Maletić, Jelena Molnar Jazić, Malcolm Watson, Jelena Tričković and Jasmina Agbaba

Department of Chemistry, Biochemistry and Environmental Protection, Faculty of Sciences, University of Novi Sad, Trg Dositeja Obradovića 3, 21000 Novi Sad, Serbia; aleksandra.tubic@dh.uns.ac.rs (A.T.); snezana.maletic@dh.uns.ac.rs (S.M.); jelena.molnar@dh.uns.ac.rs (J.M.J.); malcolm.watson@dh.uns.ac.rs (M.W.); jelena.trickovic@dh.uns.ac.rs (J.T.); jasmina.agbaba@dh.uns.ac.rs (J.A.)
* Correspondence: maja.loncarski@dh.uns.ac.rs; Tel.: +381-21-485-2798

Received: 2 October 2019; Accepted: 7 November 2019; Published: 10 November 2019

Abstract: Microplastics and chlorinated phenols (CPs) are pollutants found ubiquitously in freshwater systems. Meanwhile, bioplastics are attracting much attention as alternatives to conventional plastics, but there is little data about their effect on the behaviour of pollutants. This work therefore investigates the sorption of four CPs (4-chlorophenol—4-CP, 2,4-dichlorophenol—2,4-DCP, 2,4,6-trichlorophenol—2,4,6-TCP and pentachlorophenol—PCP) on three different plastics (polyethylene (PEg), polypropylene (PP) and polylactic acid (PLA)) using kinetics and isotherm studies. All experiments were carried out in a synthetic water matrix and in spiked Danube river water. In all cases, adsorption kinetics fitted well with the pseudo-second order rate model. Adsorption proceeded through two linear phases, corresponding to transport from the bulk solution to the external surfaces and then into the interior pores of the sorbents. Maximum adsorption capacities calculated with the Langmuir isotherm indicated that whereas adsorption of 4-CP was not significantly affected by the type of plastic present, the adsorption of 2,4-DCP, 2,4,6-TCP and PCP varied greatly, with polypropylene showing the greatest affinity for CPs adsorption. The differences observed between the adsorption behaviour of CPs in the synthetic and natural water matrices suggest further investigation is required into how the different fractions of natural organic matter impact interactions between CPs and plastics.

Keywords: microplastic; bioplastic; chlorinated phenols; sorption; kinetics; matrix effect

---

1. Introduction

It is well established that water environments have been contaminated by various pollutants, including chlorinated phenols and plastics [1–4].

Chlorophenols are ubiquitous contaminants in the environment originating from various anthropogenic activities such are chemical, textile and pharmaceutical activities [1,5]. Chlorophenols include widely applied pesticides and their degradation products. Thus, pentachlorophenol is a herbicide and insecticide used in various industries [5] and 2,4-dichlorophenol and 2,4,6-trichlorophenol are transformation products of the pesticide Triclosan [2]. Chlorophenols can also be present in drinking water due to the reaction between organic matter left in treated water and chlorine during the process of disinfection [2,5]. Furthermore, it has been confirmed that chlorophenols can be formed during disinfection, which raises particular concerns [2,6]. The European Union (EU) and United States Environmental Protection Agency (USEPA) [7–9] have therefore defined certain chlorinated phenols as priority contaminants, including 4-chlorophenol (4-CP), 2,4-dichlorophenol (2,4-DCP), 2,4,6-trichlorophenol (2,4,6-TCP) and pentachlorophenol (PCP).

It has been determined that rivers are the most at risk of all water bodies, where chlorophenols have been detected at concentrations in the range of 2–2000 µg/L [5]. Additionally, since rivers are highly connected to groundwaters, especialy by bank filtration, the presence of chlorinated phenols can pose a significant issue for drinking water treatment, not only in surface drinking water sources, but also in shallow groundwaters.

In addition to broad data indicating the adverse effects of organic pollutants to water environments and human health, in recent years plastic contamination has also attracted a lot of attention in the scientific community [3,4,10,11]. The presence of microplastics in freshwater systems is, in most cases, the consequence of inadequate communal and industrial waste management, as well as the discharge of both treated and untreated wastewater. This is becoming a major issue not only because of their direct influence on water biota, but also because many surface waters are actually drinking water sources. Polyethylene (PE) and polypropylene (PP) have been reported to represent more than 90% of the microplastics detected in drinking water sources and surface waters [4]. Recent studies have highlighted that these plastics are also frequently used as construction materials in drinking water treatment facilities, which can contribute further to the presence of these microplastics in drinking water [12]. Matsuzawa et al. (2010) [13] reported that some types of biodegradable polymers can adsorb certain hydrophobic organic compounds, including chlorinated phenols. Bioplastics have attracted much attention as alternatives of conventional plastics. Nowadays, following new EU legislation banning plastics for single use consumption, it is of special interest to investigate whether biodegradable plastics as well as petrochemical plastics can adsorb organic pollutants.

Data relating to the influence of microplastics (MP) on the marine environment and their toxicological effects on marine organisms have been thoroughly assessed. On the other hand, data on the influence of microplastics on inland water bodies and especially on drinking water sources are still very limited [4,12]. Consequently, the interactions of MPs with other pollutants that can be found in water sources, such as chlorinated phenols, are not yet fully understood. Novotna et al. (2019) [4] even suggest that some drinking water treatment plants will have to deal with microplastics as a "new" polluting agent.

The goal of this research is therefore to increase understanding of the challenges priority substances and MPs pose to groundwater treatment, by evaluating how chlorinated phenols are adsorbed both on PE and PP, the most abundant microplastics in drinking waters, and to compare these plastics with bioplastics.

2. Materials and Methods

2.1. Materials

In this study, granulated materials were used as sorbents. Two kinds of microplastic (MP) particles, polyethylene standard (PEg) and polypropylene (PP), and one bioplastic (polylactic acid, PLA), all manufactured by Sigma-Aldrich, were used. The basic physico-chemical properties of PEg of the investigated MPs and bioplastic PLA are given in Table 1.

Table 1. Physico-chemical properties of the investigated polymers.

Compound	Particle size (mm)	Density (g/cm^3)	Crystal-linity (%)	Melting Temp. (°C)	Glass Transition Temp. (°C)	Reference
PEg	3.0 *	0.918 *	44.0	114	−120	[14]
PP	3.0 *	0.9 *	38.0	165	−18	[15,16]
PLA	3.0 *	1.24 *	20.9	173–178	60–65	[17,18]

Polyethylene (PEg), polypropylene (PP), polylactic acid (PLA)*Provided by the supplier.

Four chlorinated phenols (purchased from Pestanal® Sigma-Aldrich), 4-chlorophenol (4-CP), 2,4-dichlorophenol (2,4-DCP), 2,4,6-trichlorophenol (2,4,6-TCP) and pentachlorophenol (PCP) were investigated in this work. The physico-chemical properties of the CPs are presented in Table 2.

Table 2. Physico-chemical properties of the investigated chlorophenols.

Compound	MW	$\log K_{ow}$ [a]	V_i [a]	S_w [a]	pK_a [a]
4-CP	129	2.40	1.02	27,100	9.41
2,4-DCP	163	3.06	1.14	4500	7.90
2,4,6-TCP	197	3.69	1.26	800	6.40
PCP	266	5.12	1.39	14	4.80

Pentachlorophenol (PCP). *MW*, molecular weight (g/mol); K_{ow}, octanol-water partition coefficient; V_i, McGowan volume in units of $(cm^3/mol)/100$; S_w, water solubility (mg/L); pK_a dissociation constant. [a] Kragulj et al. (2013).

These CPs differ in hydrophobicity (octanol–water partition coefficient, $\log K_{ow}$), water solubility (S_w) and acid dissociation constants (pK_a).

Hexane and methanol were purchased from J.T.Baker (for organic residue analysis), acetic anhydride and hydrogen peroxide from Sigma-Aldrich. Analytical grade reagents, anhydrous calcium chloride ($CaCl_2$), sodium hydrogen carbonate ($NaHCO_3$), and magnesium sulphate heptahydrate ($MgSO_4 \cdot 7H_2O$), were also purchased from Sigma-Aldrich.

Sorption experiments were run using a batch equilibrium method. Adsorption kinetics and isotherms were investigated in two matrices. One matrix was synthetic water containing three salts ($CaCl_2$, $NaHCO_3$ and $MgSO_4 \cdot 7H_2O$), and the other was Danube river water, since it is polluted with plastics [19] and connected to groundwater drinking water sources obtained by riverbank filtration. The pH of the synthetic matrix was 7.23 ± 0.06 and thus did not require further adjustment to be comparable with the Danube river water (pH 7.45 ± 0.07). The characteristics of the investigated matrices are given in Table 3.

Table 3. Characteristics of the synthetic water and Danube river water.

Parameter	Synthetic Water	Danube River Water
pH	7.23 ± 0.06	7.45 ± 0.07
Electro conductivity 25 °C (µS/cm)	226 ± 23	333 ± 7.0
Dissolved organic carbon (mg/L)	<0.5	2.84 ± 0.12
Chloride concentration (mg/L)	52.1 ± 3.59	44.0 ± 1.52
Sulphate concentration (mg/L)	21.2 ± 4.89	25.5 ± 3.18
Hydrogen carbonate concentration (mg/L)	134 ± 6	218 ± 43

2.2. Sorption Experiments

All kinetics experiments were conducted in 30 mL glass vials at room temperature (25 °C) using 20 mg of investigated adsorbents, which were added to the synthetic matrix or freshwater. Stock solutions of all investigated CPs (1000 µg/mL) were prepared in MeOH (J.T. Baker, for organic residue analysis). The initial concentration of CPs in the experiments was 100 µg/L. The vials were sealed and placed on a digital shaker at a speed of 150 rpm (IKA® Orbital shaker KS 501 Digital). All experiments were performed in triplicate. Samples were collected at specified time intervals (2, 4, 6, 12, 24, 48, 72 and 96 h) and filtered through a 0.45 µm membrane filter. Filtered samples were prepared for gas chromatographic analysis. The obtained experimental data were fitted with three kinetic models: the pseudo-first order, pseudo-second order and Weber–Morris models.

Adsorption isotherm experiments were carried out at CP concentrations in the range of 0–100 µg/L (1, 25, 50, 75, 100 µg/L). All experiments were carried out at the pH of Danube river water (pH = 7.45 ± 0.07). After being continuously agitated for 48 h (equilibrium time), samples were collected to quantify the equilibrium concentrations of CPs in the aqueous phase. The Freundlich and Langmuir adsorption models (see Li et al. (2018) for the formulae applied) were used to fit the adsorption isotherm data [20].

2.3. Analytical Procedure, Quality Assurance and Quality Control

Determination of the selected chlorinated phenols in water was performed using gas chromatography with mass spectrometry (Agilent Technologies, 7890A GC System/5975C VL MSD) after derivatization and liquid–liquid extraction with hexane. Blank and control experiments were performed with the sorption experiments. Blank tests, containing the same amounts of water matrix and solid particles as the samples, but without the addition of chlorinated phenols, were carried out using conditions similar to those described previously, and no target compounds were found. Control tests were carried out in 20 mL of water matrix containing the same gradient of CP concentrations as the samples, but without solid particles, in order to evaluate the loss of CP resulting from additional removal processes, such as volatilization and/or sorption to the walls of the glass bottles. The method detection limits (MDLs) of the applied analytical methods ranged from 0.11 to 0.53 µg/L. The correlation coefficient for the chlorinated phenols calibration curve was higher than 0.99. All the reported concentrations of CP were corrected with the recovery efficiency and internal standards.

3. Results

3.1. Sorption Kinetics

The results of the sorption kinetics of chlorinated phenols on the PEg, PP and PLA particles in the synthetic matrix (PEg-S, PP-S, PLA-S) and Danube river water (PEg-D, PP-D, PLA-D) are shown in Figure 1. The adsorption of CPs onto plastics reached equilibrium after 48 h for all the sorbate–sorbent combinations investigated. PP exhibited the highest uptake for the compounds (2,4-DCP, 2,4,6-TCP and PCP) with higher hydrophobicity (pKow > 2.50), and pKa values near or lower than pH of the water matrices.

The uptake of chlorinated phenols by the two kinds of microplastics and bioplastic PLA increases with time until sorption equilibrium is achieved after 48 h (Figure 1). These results are in the range of the usually reported periods for gaining sorption equilibrium of the organic compounds on various types of MPs [10,21]. PP adsorbs the highest amount of 2,4-DCP, 2,4,6-TCP and PCP (from 126–144 µg/g), while 4-CP is best adsorbed on PLA (85–101 µg/g). Matrix effects are evident in the behaviour of all the MPs, and are the most pronounced for PEg, which adsorbed the lowest amounts of all the CPs in the synthetic matrix (up to 65.9 µg/g for 4-CP, 104 µg/g for 2,4-DCP, 99.5 µg/g for 2,4,6-TCP and 53.7 µg/g for PCP), while the amounts adsorbed in the Danube river water were much higher (up to 77.7 µg/g for 4-CP, 134 µg/g for 2,4-DCP, 116 µg/g for 2,4,6-TCP and 88.0 µg/g for PCP). The same trend, but with much lower differences between the amounts of CPs adsorbed from the synthetic and Danube river water matrices, can be observed for PLA. The lowest matrix effect was observed for PP.

The kinetics data were fit to the pseudo-first order and pseudo-second order kinetic models. Correlation coefficients (R^2) for the first-order kinetics models were in the range of 0.10 for sorption of PCP on PLA to 0.95 for sorption of 2,4,6-TCP on PP_D, and in general were very low, suggesting that the pseudo-first-order model is not a good fit for the experimental data. In contrast, R^2 values for the pseudo-second order model were very high (see Figure 2). The R^2 values were greater than 0.99 for the sorption of 2,4-DCP, 2,4,6-TCP and PCP on all three materials. Slightly lower R^2 values were obtained for 4-CP, but the values are still high, ranging from 0.9460 to 0.9890. The calculated q_e, k_2, and R^2 values for the pseudo-second order model are listed in Table 4. The q_e values obtained agree well with the experimentally measured q_e values. Based on these results, the pseudo-second order equation can successfully be used to describe the sorption process of the four investigated chlorinated phenols onto PEg, PP and PLA particles, from the beginning of the sorption process to the equilibrium stage. This means that chlorinated phenols can be adsorbed to different binding sites on the investigated MPs and PLA [11].

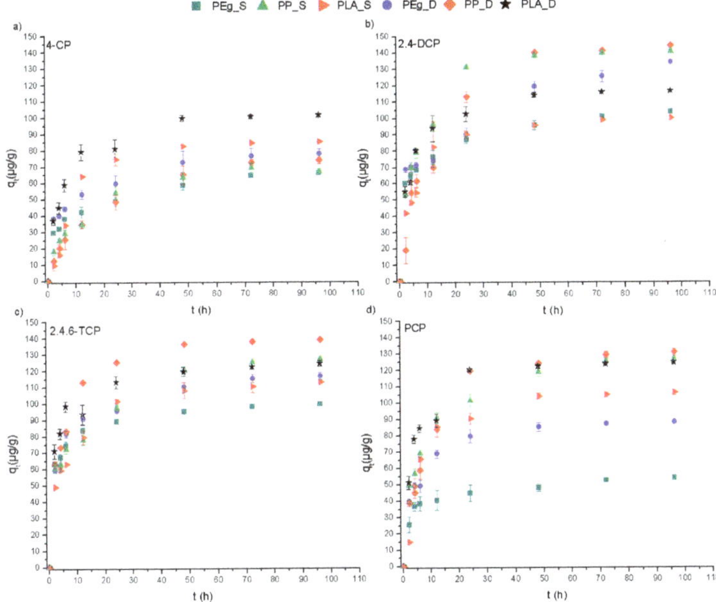

Figure 1. Experimental data ($n = 3$, mean value ± SD) of (**a**) 4-CP; (**b**) 2,4-DCP; (**c**) 2,4,6-TCP and (**d**) PCP on PEg, PP and PLA particles in the synthetic matrix (PEg_S, PP_S and PLA_S) and Danube river water (PEg_D, PP_D and PLA_D).

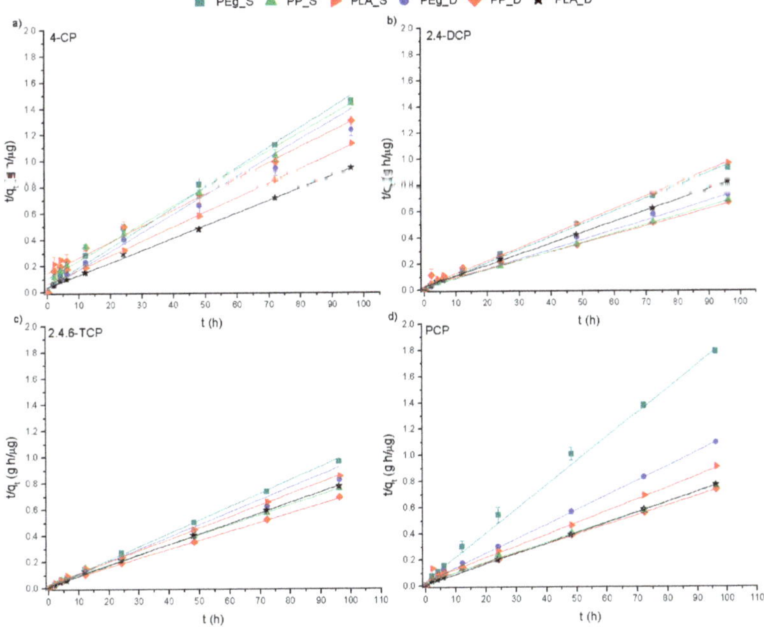

Figure 2. Plots for the sorption kinetics, based on the pseudo-second order model, of (**a**) 4-CP; (**b**) 2,4-DCP; (**c**) 2,4,6-TCP and (**d**) PCP on PEg, PP and PLA particles in the synthetic matrix (PEg_S, PP_S and PLA_S) and Danube river water (PEg_D, PP_D and PLA_D) ($n = 3$, mean value ± SD).

Table 4. Parameters calculated with pseudo-second order kinetic models for the sorption of chlorinated phenols onto PEg, PP and PLA particles in synthetic and Danube river water.

Compounds	Solid Phase	k_2 (h^{-1})	R^2	q_e (Theoretical)	q_e (Experimental)
4-CP	PEg_S	0.0063	0.9890	65.85	65.49
	PEg_D	0.0065	0.9460	77.69	70.22
	PP_S	0.0018	0.9749	66.85	71.84
	PP_D	0.0010	0.9815	73.62	83.13
	PLA_S	0.0019	0.9939	85.02	90.83
	PLA_D	0.0025	0.9981	101.4	105.6
2,4-DCP	PEg_S	0.0059	0.9912	103.7	102.2
	PEg_D	0.0032	0.9925	133.8	135.3
	PP_S	0.0022	0.9978	140.4	145.8
	PP_D	0.0012	0.9915	143.8	153.8
	PLA_S	0.0031	0.9977	99.62	101.8
	PLA_D	0.0026	0.9993	116.4	120.6
2,4,6-TCP	PEg_S	0.0085	0.9973	99.56	98.14
	PEg_D	0.0056	0.9930	116.5	105.8
	PP_S	0.0022	0.9974	126.9	130.7
	PP_D	0.0020	0.9991	138.9	145.6
	PLA_S	0.0026	0.9937	112.9	115.6
	PLA_D	0.0046	0.9992	124.1	125.2
PCP	PEg_S	0.0087	0.9990	53.68	54.38
	PEg_D	0.0041	0.9985	88.05	90.09
	PP_S	0.0021	0.9952	126.3	129.9
	PP_D	0.0017	0.9935	130.5	136.8
	PLA_S	0.0020	0.9789	105.9	112.1
	PLA_D	0.0060	0.9988	124.5	126.7

Differences were observed between the most hydrophilic 4-CP and the most hydrophobic PCP on PEg, PP and PLA, with adsorptions ranging from 65.9 to 101 µg/g to 53.7 to 126 µg/g, respectively.

The plots obtained using the Weber–Morris intraparticle diffusion model (shown in Figure 3) imply that the sorption of all four chlorophenols on PEg, PP and PLA proceeds through two linear phases, corresponding to the transport of the sorbate from the bulk solution to the external surfaces and then into the interior pores of the sorbent [10,22]. For most combinations of CPs and MPs/bioplastics, adsorption at the beginning is fast for the first five hours. After that period, a slower adsorption process is observed. The slowest adsorption rate was observed for all four CPs on PEg in both water matrices. The linear plots obtained by the Weber–Morris model do not pass through the origin for the sorption of all four chlorinated phenols on all three materials, which indicates that surface sorption could be the rate-controlling step [11].

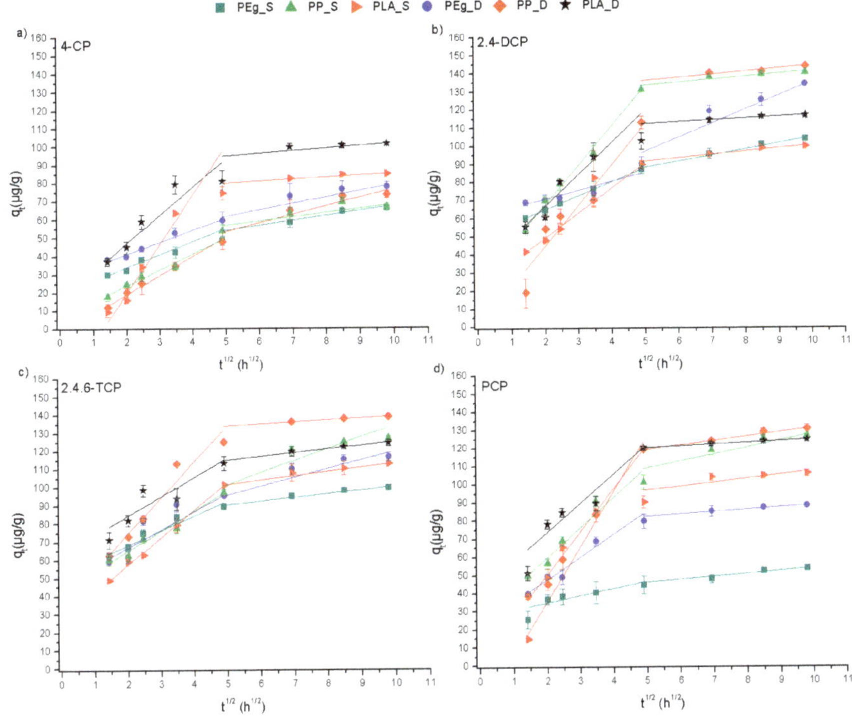

Figure 3. Plots for the sorption kinetics, based on the Weber–Morris model, of (**a**) 4- CP; (**b**) 2,4-DCP; (**c**) 2,4,6-TCP and (**d**) PCP on PEg, PP and PLA particles in the synthetic matrix (PEg_S, PP_S and PLA_S) and Danube river water (PEg_D, PP_D and PLA_D) ($n = 3$, mean value ± SD).

3.2. Adsorption Isotherms

In order to investigate the sorption mechanism of the chlorinated phenols on microplastics and bioplastics in the two matrices investigated, the Freundlich and Langmuir models were applied. These are the two most commonly applied models for describing the mechanism of interactions between sorbents and sorbates. The Freundlich model is applicable to both monolayer and multilayer sorption and assumes that the sorption process takes place on the heterogeneous surface of a sorbent [10,23]. The application of the Langmuir model can show whether the sorbate makes a monolayer coverage on homogenous sorbent surfaces [1,10]. Figure 4 presents the adsorption isotherms of chlorinated phenols on PEg, PP and PLA in the synthetic and Danube river water matrices. The values of the Freundlich and Langmuir model parameters are listed in Table 5.

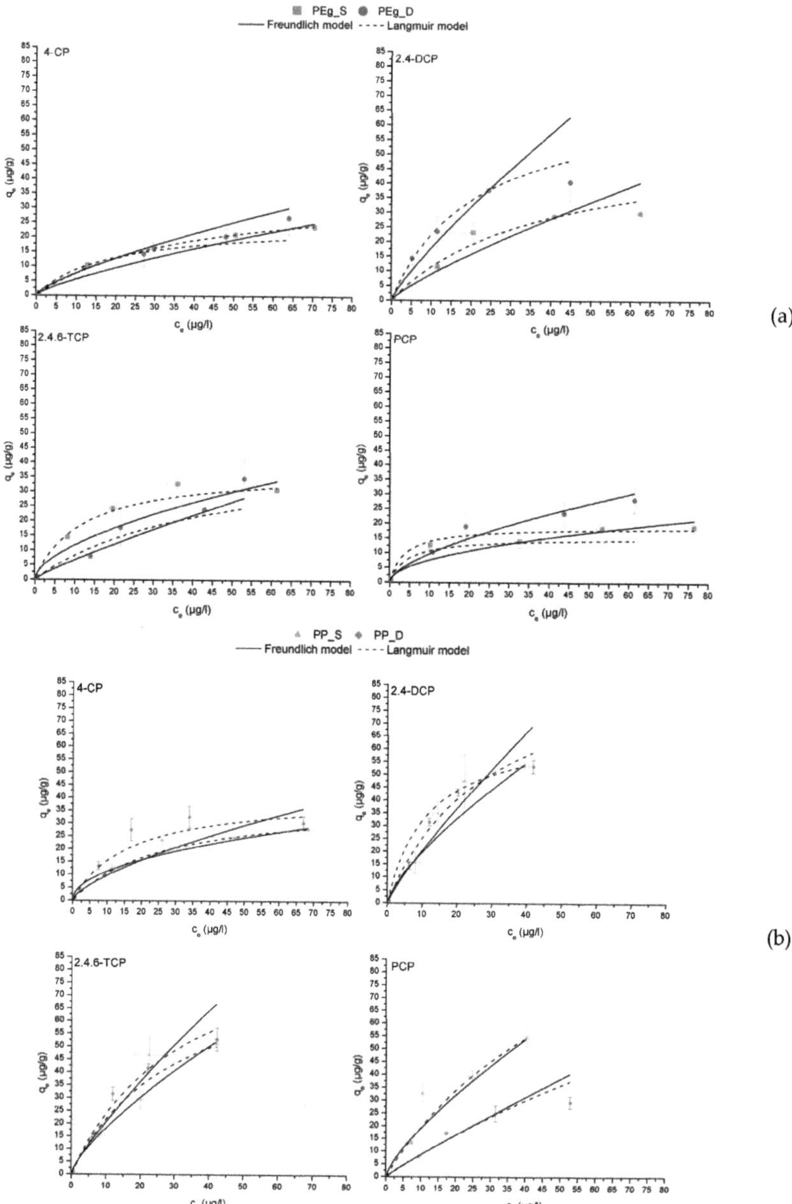

Figure 4. *Cont.*

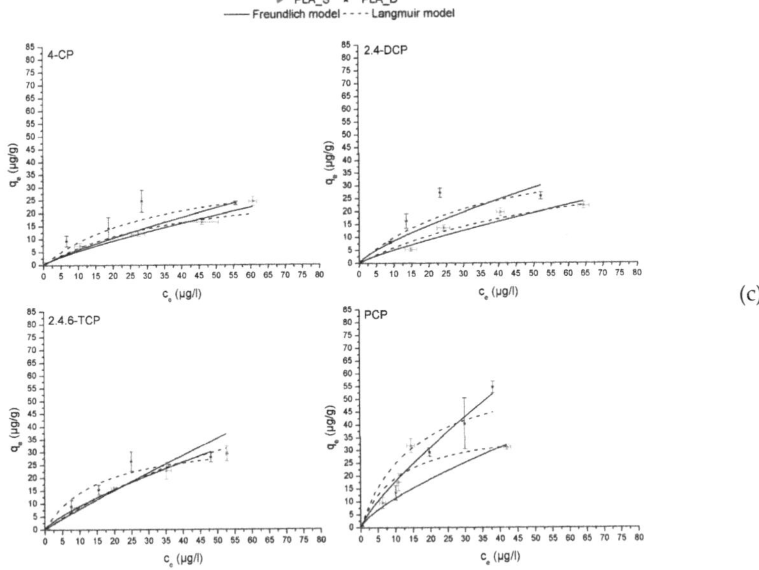

Figure 4. Sorption isotherms of 4-CP, 2,4-DCP, 2,4,6-TCP and PCP on (a) PEg, (b) PP and (c) PLA particles obtained for the synthetic matrix (PEg_S, PP_S and PLA_S) and Danube river water (PEg_D, PP_D and PLA_D) (n = 3, mean value ± SD).

Table 5. Values of the Freundlich and Langmuir model parameters for the sorption of 4-CP, 2,4-DCP, 2,4,6-TCP and PCP on PEg, PP and PLA.

Compounds	Adsorbents	Freundlich model			Langmuir model		
		R^2	n	$K_F(\mu g/g)/(\mu g/L)^n$	R^2	$q_{max}(\mu g/g)$	$K_L(l/\mu g)$
4-CP	PEg_S	0.9478	0.77	0.92	0.9892	37.03	0.0255
	PEg_D	0.9816	0.74	1.36	0.9825	24.55	0.0551
	PP_S	0.9974	0.48	3.69	0.9984	39.17	0.0365
	PP_D	0.8492	0.69	1.90	0.9687	41.15	0.0599
	PLA_S	0.9869	0.80	0.85	0.9762	32.74	0.0252
	PLA_D	0.9835	0.82	0.91	0.9938	38.40	0.0298
2,4-DCP	PEg_S	0.9949	0.83	1.33	0.9989	55.60	0.0258
	PEg_D	0.9998	0.84	2.58	0.9998	71.36	0.0457
	PP_S	0.9970	0.74	3.63	0.9945	68.98	0.0869
	PP_D	0.9347	0.85	2.86	0.9884	100.2	0.0338
	PLA_S	0.9585	0.83	0.74	0.9423	45.48	0.0152
	PLA_D	0.9268	0.76	1.52	0.9392	46.35	0.0275
2,4,6-TCP	PEg_S	0.9397	0.59	3.02	0.9390	37.47	0.0839
	PEg_D	0.9897	0.86	0.90	0.9919	44.95	0.0222
	PP_S	0.9960	0.73	3.37	0.9631	87.13	0.0330
	PP_D	0.9599	0.82	3.14	0.9911	100.1	0.0313
	PLA_S	0.9703	0.93	0.95	0.9881	82.23	0.0118
	PLA_D	0.9668	0.76	1.59	0.9796	35.59	0.0662
PCP	PEg_S	0.9521	0.51	2.32	0.9913	18.93	0.2681
	PEg_D	0.9958	0.63	2.24	0.9349	15.32	0.2261
	PP_S	0.9708	0.75	3.39	0.9438	135.2	0.0167
	PP_D	0.9884	0.93	1.02	0.9898	168.2	0.0054
	PLA_S	0.9531	0.73	2.15	0.9817	62.74	0.0940
	PLA_D	0.9925	0.80	2.88	0.9246	39.54	0.0658

The correlation coefficients for the Freundlich and Langmuir models are, respectively, $R^2 = 0.8292$–0.9998 and $R^2 = 0.9246$–0.9998 for all four CPs on all of the polymers. Significant differences between these two models were not observed.

In the Freundlich isotherm, the parameter n is an indicator of sorption linearity, with larger deviations from $n = 1$ indicating greater nonlinearity. All the isotherms obtained are nonlinear (n values ranging from 0.48 to 0.93), suggesting heterogenic binding sites on the surface of the sorbents (Table 5). In general, greater nonlinearity was observed in the synthetic matrix than in the Danube river water, demonstrating the influence of this matrix on the sorption process.

In the case of the Langmuir model, it is assumed that sorption of the sorbate molecule occurs at a specific site on the sorbent, with no further sorption occurring at the same site [10,23]. The q_{max} are in the ranges 24.55–41.15 µg/g, 46.35–100.2 µg/g, 35.59–100.1 µg/g and 15.3–168.2 µg/g for 4-CP, 2,4-DCP, 2,4,6-TCP and PCP on PEg, PP and PLA, respectively. These differences confirm that the structure of the compound has a significant impact on its interactions with the plastic surfaces. PP showed a considerably higher sorption affinity for CPs than PLA and PEg, which confirms that sorption is dependent on the sorbent properties. The difference between the three sorbents investigated is most pronounced regarding PCP sorption. The highest amounts of PCP ($q_{max} = 135.2$–168.2 µg/g) are adsorbed on PP, following 2,4,6-TCP, 2,4-DCP and 4-CP. The lowest q_{max} values were obtained for the sorption of PCP on PEg ($q_{max} = 15.32$–18.93 µg/g). In general, all the CPs had lower sorption affinities for PEg than either of the other two plastics investigated. Matrix influences on q_{max} can also be observed, with higher values generally observed for PEg and PP in Danube river water compared to the synthetic matrix. The adsorption capacity of PLA towards CPs in Danube river water had similar values (35.52–46.35 µg/g), and these are similar or lower than in the synthetic matrix.

4. Discussion

Plastic debris is now found throughout the environment, and may serve as a carrier for organic pollutants, influencing not just their transport through various environments, but potentially also influencing how they express their toxicity towards aquatic organisms and humans [3]. Bioplastics are also becoming more broadly applied in everyday life, increasing their discharge to the environment. Understanding the interactions between various plastics and organic pollutants is therefore extremely important [10,11,21,24].

The adsorption kinetics and isotherm data presented above demonstrate significant variations in the degree to which microplastics interact with chlorinated phenols in neutral pH aqueous environments. The different polarities of the phenols studied, as demonstrated by their various pKa values, mean that under the conditions investigated some of these compounds are partially present in disassociated forms, rather than as neutral species [11,13]. The different plastics have differing monomeric compositions, with the linear aliphatic PE, and the slightly branched aliphatic PP and PLA which is an aliphatic polyester. They are therefore likely to undergo different interactions with non-hydrophobic compounds such as phenols, as suggested by other authors [10,25]. The obtained results indicate that the structure of microplastic is significant for sorption processes when the available surfaces of the polymers are similar, which is in accordance with the finding of other authors [24,25]. PP showed the highest sorption capacity towards all four CPs, followed by PLA and PEg, suggesting that nonlinear polymers exhibit a higher potential to adsorb and transport ionisable compounds through the environment. One can assume that the hydrophobic character of these compounds is responsible for the majority of interactions between CPs and PP, because with increases in compound hydrophobicity, in both water matrices increases in the adsorption capacity were observed. The poor sorption of ionised PCP and 2,4,6-TCP observed on PEg are probably explained by electrostatic repulsion (Table 5). The best sorption on this plastic was observed for 2,4-DCP, which is hydrophobic and non-ionised under the prevailing pH conditions of the experiment. The adsorption capacity of PE towards 4-CP is similar to 2,4,6-TCP, indicating that the slightly hydrophilic nature of this compound and its non-ionised structure both influence the sorption. In the synthetic matrix, the capacity of PLA to adsorb the hydrophobic ionised molecules of 2,4,6-TCP and PCP is two times higher than for the non-ionised 4-CP and 2,4-DCP.

The presence of other water constituents in the Danube river water, especially the natural organic matter (NOM), was expected to potentially interfere with the adsorption of the CPs by competing for adsorption sites on the surface of the plastics. However, the reverse trend was observed, whereby the q_{max} of CPs in the presence of PEg, PP and PLA was generally greater or similar in the experiments carried out with Danube river water, in comparison to the synthetic matrix. This could indicate that additional interactions with the NOM were present. The adsorption of 2,4,6-TCP and PCP on PLA were the only cases where the q_{max} values were significantly lower in the presence of NOM. NOM itself is a complex mixture of organic substances with various properties and functional groups. Earlier characterisation of the Danube river NOM [26] revealed that the majority of NOM present is mainly of hydrophilic character, while the hydrophobic fraction is relatively insignificant. This topic will certainly be of interest for future research, as also indicated by Xu et al. [21,25].

5. Conclusions

The results obtained in this work demonstrate that the presence of microplastics and bioplastics in aquatic environments may have a significant impact on the fate and transport of certain chlorinated phenols.

The adsorption of 4-CP was not significantly affected by the type of plastic present or the water matrix. In contrast, the type of plastic had a great impact on adsorption of 2,4-DCP, 2,4,6-TCP and PCP, with PCP demonstrating the most pronounced variation. In general, polypropylene showed the greatest affinity for the adsorption of CPs, whereas adsorption onto polyethylene and polylactic acid was more dependent on the structure of the compounds.

The differences observed between the adsorption behaviour of CPs in the investigated synthetic and natural water matrices demonstrate the need for further investigation, particularly in terms of how the role of different fractions of natural organic matter impact these interactions.

Author Contributions: conceptualization, A.T., J.A., S.M.; methodology, A.T., M.L.; investigation, M.L.; resources, J.A., J.T.; writing—original draft preparation, A.T., M.L.; writing—review and editing, J.M.J., M.W.; Visualization, A.T. and M.L.; project administration, J.T.; funding acquisition, J.A., J.T.

Funding: This research was funded by the Ministry of Education, Science and Technological Development of the Republic of Serbia (Project No. III43005).

Acknowledgments: The authors gratefully acknowledge the support of the Ministry of Education, Science and Technological Development of the Republic of Serbia (Project No. III43005).

Conflicts of Interest: The authors declare no conflict of interest.

References

1. Czaplicka, M. Sources and transformations of chlorophenols in the natural environment. *Sci. Total. Environ.* **2004**, *322*, 21–39. [CrossRef] [PubMed]
2. Vikesland, P.J.; Fiss, E.M.; Wigginton, K.R.; McNeill, K.; Arnold, W.A. Halogenation of Bisphenol-A, Triclosan, and Phenols in Chlorinated Waters Containing Iodide. *Environ. Sci. Technol.* **2013**, *47*, 6764–6772. [CrossRef] [PubMed]
3. Koelmans, A.A.; Nor, N.H.M.; Hermsen, E.; Kooi, M.; Mintenig, S.M.; De France, J. Microplastics in freshwaters and drinking water: Critical review and assessment of data quality. *Water Res.* **2019**, *155*, 410–422. [CrossRef] [PubMed]
4. Novotna, K.; Cermakova, L.; Pivokonska, L.; Cajthaml, T.; Pivokonsky, M. Microplastics in drinking water treatment-Current knowledge and research needs-review. *Sci. Total. Environ.* **2019**, *667*, 730–740. [CrossRef] [PubMed]
5. Michałowicz, J.; Duda, W. Phenols-sources and toxicity. *Pol. J. Environ. Stud.* **2007**, *16*, 347–362.
6. Gallard, H.; von Gunten, U. Chlorination of Phenols: Kinetics and Formation of Chloroform. *Environ. Sci. Technol.* **2002**, *36*, 884–890. [CrossRef]

7. The European Parliament and the Council of the European Union. *Directive 2013/39/EU of the European Parliament and of the Council of 12 August 2013 Amending Directives 2000/60/EC and 2008/105/EC as Regards Priority Substances in the Field of Water Policy Text with EEA Relevance: Directives 2000/60/EC and 2008/105/EC as Regards Priority Substances in the Field of Water Policy*; The European Parliament and the Council of the European Union: Brussels, Belgium, 2013.

8. The European Parliament and the Council of the European Union. *Directive 2008/105/EC of the European Parliament and of the Council of 16 December 2008 on Environmental Quality Standards in the Field of Water Policy, Amending and Subsequently Repealing Council Directives 82/176/EEC, 83/513/EEC, 84/156/EEC, 84/491/EEC, 86/280/EEC and Amending Directive 2000/60/EC of the European Parliament and of the Council: Directive on EQS in the Field of Water Policy*; The European Parliament and the Council of the European Union: Strasbourg, France, 2008.

9. U.S. Environmental Protection Agency (EPA). *2004 Edition of the Drinking Water Standards and Health Advisories*; EPA: Washington, DC, USA, 2004.

10. Wang, W.; Wang, J. Comparative evaluation of sorption kinetics and isotherms of pyrene onto microplastics. *Chemosphere* **2018**, *193*, 567–573. [CrossRef]

11. Guo, X.; Pang, J.; Chen, S.; Jia, H. Sorption properties of tylosin on four different microplastics. *Chemosphere* **2018**, *209*, 240–245. [CrossRef]

12. Mintenig, S.M.; Löder, M.G.J.; Primpke, S.; Gerdts, G. Low numbers of microplastics detected in drinking water from ground water sources. *Sci. Total. Environ.* **2019**, *648*, 631–635. [CrossRef]

13. Matsuzawa, Y.; Kimura, Z.-I.; Nishimura, Y.; Shibayama, M.; Hiraishi, A. Removal of Hydrophobic Organic Contaminants from Aqueous Solutions by Sorption onto Biodegradable Polyesters. *JWARP* **2010**, *2*, 214–221.

14. Paszkiewicz, S.; Szymczyk, A.; Pawlikowska, D.; Subocz, J.; Zenker, M.; Masztak, R. Electrically and Thermally Conductive Low Density Polyethylene-Based Nanocomposites Reinforced by MWCNT or HybridMWCNT/Graphene Nanoplatelets with Improved Thermo-Oxidative Stability. *Nanomaterials* **2018**, *8*, 264. [CrossRef] [PubMed]

15. Maddah, H.A. Polypropylene as a Promising Plastic: A Review. *AJOP* **2016**, *6*, 1–11.

16. Mandolfino, C. Polypropylene surface modification by low pressure plasma to increase adhesive bonding: Effect of process parameters. *Surf. Coat. Tech.* **2019**, *366*, 331–337. [CrossRef]

17. Middleton, J.C.; Tipton, A.J. Synthetic biodegradable polymers as orthopedic devices. *Biomaterials* **2000**, *21*, 2335–2346. [CrossRef]

18. Kaavessina, M.; Ali, I.; Al-Zahrani, S.M. The Influences of Elastomer toward Crystallization of Poly (lactic acid). *Procedia Chem.* **2012**, *4*, 164–171. [CrossRef]

19. Liedermann, M.; Gmeiner, P.; Pessenlehner, S.; Haimann, M.; Hohenblum, P.; Habersack, H. A Methodology for Measuring Microplastic Transport in Large or Medium Rivers. *Water* **2018**, *10*, 414. [CrossRef]

20. Li, J.; Zhang, K.; Zhang, H. Adsorption of antibiotics on microplastics. *Environ. Pollut.* **2018**, *237*, 460–467. [CrossRef]

21. Xu, B.; Liu, F.; Brookes, P.C.; Xu, J. The sorption kinetics and isotherms of sulfamethoxazole with polyethylene microplastics. *Mar. Pollut. Bull.* **2018**, *131*, 191–196. [CrossRef]

22. Weber, T.W.; Chakravorti, R.K. Pore and solid diffusion models for fixed bed adsorbers. *AICHE J.* **1974**, *20*, 228–238. [CrossRef]

23. Foo, K.Y.; Hameed, B.H. Insights into Modeling of Adsorption Isotherm Systems. *Chem. Eng. J.* **2010**, *156*, 2–10. [CrossRef]

24. Huffer, T.; Hofmann, T. Sorption of non-polar organic compounds by micro-sized plastic particles in aqueous solution. *Environ. Pollut.* **2016**, *214*, 194–201. [CrossRef] [PubMed]

25. Xu, B.; Liu, F.; Brookes, P.C.; Xu, J. Microplastics play a minor role in tetracycline sorption in the presence of dissolved organic matter. *Environ. Pollut.* **2018**, *240*, 87–94. [CrossRef] [PubMed]

26. Tubić, A.; Leovac, A.; Molnar, J.; Krčmar, D.; Paunović, O.; Ivančev-Tumbas, I. Characterization of Dissolved Organic Matter from the Danube River Before and After Ozone Oxidation. In Proceedings of the 6th Symposium Chemistry and Environmental Protection EnviroChem, Vršac, Serbia, 21–24 May 2013; pp. 282–283.

MDPI

St. Alban-Anlage 66

4052 Basel

Switzerland

Tel. +41 61 683 77 34

Fax +41 61 302 89 18

www.mdpi.com

Water Editorial Office

E-mail: water@mdpi.com

www.mdpi.com/journal/water